Analysis
of
Vertebrate Populations

Analysis
of
Vertebrate Populations

GRAEME CAUGHLEY

Senior Lecturer in Biology
School of Biological Sciences
The University of Sydney

A Wiley–Interscience Publication

JOHN WILEY & SONS

Chichester · New York · Brisbane · Toronto

Library of Congress Cataloging in Publication Data:

Caughley, Graeme.
  Analysis of vertebrate populations.

  'A Wiley–Interscience publication.'
  Bibliography: p.
  1. Animal populations. I. Title. [DNLM:
1. Population dynamics. 2. Population surveillance.
3. Vertebrates. QL752 C371a]
QH352.C38   596'.05'24   76–913
ISBN 0 471 01705 1

Photosetting by Thomson Press (India) Limited, New Delhi
Printed at The Pitman Press, Bath

Preface

The analyses needed to penetrate a vertebrate population and to find out what makes it tick are scattered amongst several thousand books and scientific journals. The very diffuseness of this literature has often slowed my own research, particularly when the nearest library is in the next town or in the next country. Other ecologists have suffered equally. My students in applied population ecology and wildlife management also have a rough time of it because there is no single text-book that covers the ground.

This book, presenting an introduction to the field, is an attempt at a remedy. I have selected from the literature those analyses and those ways of approaching a problem that I think are immediately relevant. The gaps left in this array have been plugged temporarily with makeshift analyses of my own. Some analyses, although mathematically impeccable, require information that is difficult or impossible to collect in the field. Others are anchored on assumptions that have little or nothing to do with the realities of living things. These do not appear here. The methods have been selected with care. I am an ecologist, not a mathematician, and I have chosen those methods most easily understood and applied. Most have been used in my own work, and consequently I have some appreciation of their practical strengths and weaknesses.

Any book on population analysis is a compromise between precision and comprehensibility. When one or other had to be sacrificed, I sacrificed precision. Because many readers will not be at ease with calculus and matrix algebra these powerful techniques are not used, the only expertise assumed in the reader being a knowledge of elementary algebra and statistics.

The book has four specific aims:

1. to provide a student of ecology or wildlife management with enough of the theory of population dynamics to enable him to think in terms of population processes themselves, rather than simply in terms of their effects;

2. to outline what information he requires to solve the questions of population dynamics posed by a problem in ecology or management;

3. to describe how this information can be analysed to answer these questions; and

4. to show how population analysis is used in such practical problems as reducing a population, stimulating it to increase, or taking from it a sustained yield.

Throughout, I have illustrated analyses with examples drawn from a wide range of species and environments. The immediate purpose of each is to guide the reader through the calculations, but it also has two further functions. First, the example should be of biological interest in itself, and secondly, it should channel the reader's attention through the arithmetic to where the algebraic result is converted to a biological conclusion. This translation is the most difficult step and the most important stage of population analysis.

I am immensely grateful to L. C. Birch, and the other Sydney ecologists, for the many hours of discussions that helped shape this book. Various chapters were criticized to their benefit by A. C. Hodson, J. Monro, T. Riney and J. A. Gulland. I thank my wife who, for some peculiar reason, is Judith Caughley when she works on mammals but Judith Badham when working on reptiles. She has slipped me many a theoretical insight, being too practical to record them herself.

GRAEME CAUGHLEY

School of Biological Sciences
University of Sydney
October 1975

Contents

Chapter 1

Introduction

Population analysis is concerned with the numerical attributes of a population—numbers, sex ratio, rate of increase and so on—together with the properties of the animals and the properties of the environment that determine these values. 'Population' means different things to different people. Cole (1957) gave a working definition convenient for our purposes: a population is 'a biological unit at the level of ecological integration where it is meaningful to speak of a birth rate, a death rate, a sex ratio and an age structure in describing the properties of the unit'.

The dynamics of a population can be studied at several levels of detail. A simple approach is to treat individuals as if they were identical, to express the numbers in the population as an average over several years, and to investigate why the average has this value. One step removed from this elementary approach is the study of fluctuations in total numbers as related to changes in the environment. At a more advanced level the study might be aimed at expressing the rate of change in numbers as a difference between birth rate and death rate, the difference being related to environmental influences. More detailed again is the study which discards the simplifying but unrealistic assumption that the animals in the population are identical. Such an approach recognizes that the environment does not act directly on numbers as such but indirectly through its influence on fecundity at each age and survival over each interval of age. Environmental influences would be related directly to these attributes rather than to numbers or to change in numbers. At the final stage of this progression each individual is recognized as unique. The dynamics of the population are reported as the sum of the demographic reactions of each individual. That approach may seem ideal but with it generality is lost in a porridge of detail. Between the oversimplification of declaring all animals identical, and the self-defeating undersimplification of reducing a population to its fundamental particles, there lies a broad middle ground forming the ecologist's methodological domain.

The range of approaches, from gross to detailed, allows considerable flexibility in the way a problem is attacked. These are different approaches to the one question: what are the relationships between animals and the conditions in which they live? The level at which a problem is entered depends both on how precise the answer needs to be and on the difficulty of extracting an answer of that accuracy. And it depends also on the attitude of the researcher. If he feels that general principles of ecology lie near the surface, he will be content with a gross approach. If, however, he suspects that the particular god ruling things ecological is not only subtle but frequently malign, he will investigate a problem in greater detail.

My bias is of the second kind. I am struck by the lack of progress in population ecology over the last 50 years. We cannot yet point to a single well-established principle in this field. Perhaps the answers we seek are not necessarily for the taking, and simple approaches, rather than leading to simple answers, may more often lead to no answers at all. I strongly suspect that the deepest insight into a population comes from studying how age-specific survival and fecundity are influenced by the conditions in which the animals live. Such an approach cuts deeper into a problem of population ecology than does any of the others. We need its keen edge to expose the principles of population ecology that have, so far, eluded us.

An ecologist rightly reacts against reducing his animals to the algebraic abstractions of x, y, and z, but a greater peril awaits him when he reluctantly takes this step. Algebra has a seductive beauty of its own; it can lure him away from the rough prose of population into the elegant poetry of probability. The population ecologist studies animals but he needs to translate these into algebra and to proceed to an algebraic conclusion. This is not an end-point; the conclusion must then be transformed back to animals living in a real world. The conclusion expressed in terms of the animals themselves is invariably dishevelled compared with the algebraic conclusion, but the animals rather than the mathematics are the subject of study and the conclusion must be biological, not mathematical.

A population dynamicist must, above all else, be an ecologist. He must know the animal he studies and must have some understanding of the components of the environment that influence it. The first part of any study should be spent in becoming familiar with the species. Assumptions of analysis cannot be matched to the lives of the animals without a knowledge of life history and behaviour. Common sense is the most important requirement; it holds mathematics to reality. No analysis in this book will exactly fit the needs of a particular study and modifications will be needed, but these can be made only when the assumptions underlying an analysis are fully understood. Chaos usually results from using an analysis as if it were a cake recipe.

Studies on population dynamics can be divided into three categories: the first comprises analyses of human populations and populations of insects; fish are the subject of the second, with whales tacked on as an afterthought; and in the third category are most of the studies on mammals and birds. Each category has its own notation and its own methods of analysis. The lines of demarcation are surprisingly sharp, hardly blurred at all by the few enterprising studies that ignored the boundaries. All this is decidedly odd; it suggests that those people working on one group of animals seldom read publications about other groups, the taxon of the animal apparently being considered a more important attribute than the biological process being investigated. This, of course, is nonsense. A birth rate is a birth rate whether bears or beetles are born, and the computations are the same in each case. But there is a difference—beetles can be studied in bottles: bears are not fit subjects for study in a laboratory—and the two sets of data must be collected in different ways. The population of bears may not provide data allowing direct estimation of certain parameters, indirect methods being required, and such instances call for methods specific to a particular problem of sampling. Some of these will be described, but in relation to specific problems rather than to specific taxa. Although the examples in this book are taken mainly from mammals and birds, I have applied to them several methods that were developed for other groups; and, of course, most of the methods presented here can be applied with little modification to problems involving fish, reptiles, invertebrates and plants.

Fisheries biologists may not feel at ease with some of the analyses in this book. The age-specific parameters of fecundity and mortality used here in abundance are usually difficult and often impossible to estimate for fish, and consequently seldom appear explicitly in analyses of fish populations. Researchers in this field have outflanked acute problems of sampling by using an indirect mathematical approach (see particularly Beverton and Holt 1957; Ricker 1958) of awe inspiring complexity and imaginative scope. I am unable to add to it, and reluctant to risk luring a fisheries biologist away from it, but he may find that some of the methods presented here complement those used in fisheries research. Murphy's (1967) application to sardine populations of parameters originally developed by students of human demography shows that the gap between the two approaches can be bridged.

Subsequent chapters deal with problems of population dynamics, the data needed to investigate them, and the methods by which the data are analysed to solve them. Field techniques are not touched upon, except in passing, but introductions to these methods are available elsewhere for insects (Southwood 1966), fish (Allen 1951, Rounsefell and Everhart 1953, Gerking 1967), birds (Wolfson 1955, Giles 1969), reptiles (Goin and Goin 1962) and mammals (Davis 1956, Southern 1964, Giles 1969).

Chapter 2

The population

Of the various analogies by which a population can be described, the most useful is that provided by Farner (1955). He drew attention to the close similarity between the dynamics of a population and the dynamics of a lake. Water flows into the lake (births) and out of it (deaths), and the recent history of interplay between these largely determines the level of the lake (animal numbers). To extend the analogy, the lake level is also influenced by direct precipitation (immigration) and evaporation (emigration). Populations are more complex than lakes, and the analogy would break down if elaborated further, but their basic properties are held in common. Both are open systems, and they can be described by the same equations.

2.1 LIMITS OF A POPULATION

The concept of population that we carry in our minds—a group of inter-breeding individuals having little or no contact with other such groups—is far removed from what we actually see. Figure 2.1 diagrams a much more common pattern of dispersion. It shows a mosaic distribution in which groups of animals are discernible but no group can be considered entirely discrete. Four 'populations' are enclosed by the nested rectangles. The boundaries of rectangle A bear no relationship to the boundaries of any group of animals. Here a 'population' comprises the animals on an area of arbitrary size and shape. The 'population' of mice in a field and the deer 'population' of New York State are of this kind.

Rectangle B heeds the boundary of a group of animals, but so also do rectangles C and D. Each encloses a 'population' in the sense defined in Chapter 1; convenience determines which population is chosen for study.

The important point is not that populations are difficult to define in space and time but that a research worker must have a clear idea of the relationship between the boundaries of his study area and the partial or complete disconti-nuities between aggregations of animals. Otherwise his estimates of population size, rate of increase and so on, may be referrable not only to the 'population' on his study area but to the animals in the areas peripheral to it. A mark-

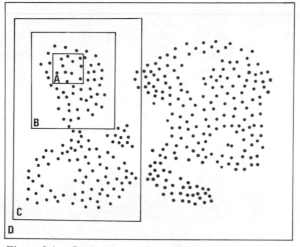

Figure 2.1. Study areas (nested rectangles) imposed
upon a population.

recapture estimate of the number of animals on rectangle A (Figure 2.1),
for instance, would return an exaggerated result because of movement within
rectangle B. This problem is minimized by making the study area as large as
circumstances permit. With increasing size the ratio of circumference to area
declines, and with it declines the percentage of animals moving across the
boundary.

2.2 BREEDING SYSTEMS

The demographic properties of a population are determined by the life-
history pattern of its members. Cole (1954) investigated the demographic
consequences of several different life histories and showed that when populations are sufficiently large, and a sufficient span of time is considered, the
overall results are similar. Thus although specific equations will predict the
demographic consequences of any single pattern of life history, general
equations are usually no less accurate.

Some life-history patterns are special cases of more general patterns. Populations have, for instance, been divided into those with overlapping generations
and those whose generations are discrete in time. Many insects are of the second
kind, the individuals taking one year to reach maturity and then dying immediately after they reproduce. Although different equations have been used
for these two cases, discrete generations are mathematically a special case of
overlapping generations, and a system of mathematics adequately describing
the second is entirely appropriate to the first.

The advantages of describing all populations in terms of one or two
models, rather than in terms of ten or a dozen, are only too obvious. The only

division recognized here, for purposes of computing statistics, is between populations that produce offspring at a rate almost constant throughout the year and those that produce offspring over a restricted season. Although these two cases can be described by the same system of equations, they are more conveniently treated according to two models. The first, a birth-flow model, describes a population whose rate of breeding is constant throughout the year. The second, a birth-pulse model, describes the demographic behaviour of a population that produces all offspring for the year on one day, the date being the same from year to year. No population exactly fits either model, but most populations come closer to one than the other. If a population tends towards constancy of reproductive rate, as do populations of man, red kangaroos, *Megaleia rufa* (Frith and Sharman 1964) and Andean sparrows, *Zonotrichia capensis* (Miller 1962), they can be described reasonably accurately by the parameters of the birth-flow model. Those whose season of births has a standard deviation less than 30 days, and most do, can be described by the birth-pulse model. The few intermediate cases can be treated both ways and parameters can be estimated as means between the two models.

Birth-pulse populations, being the commonest, are given the most space in this book but birth-flow analyses are outlined where appropriate.

2.3 CHOICE OF PARAMETERS

To understand the dynamics of a population we need to know how many animals it contains, how fast it is increasing or decreasing, its rate of production of newborn and its rate of loss through mortality. Properties such as these are called parameters; estimates of parameters are called statistics. A population can be described by an infinite number of parameters but for most purposes a small number is sufficient. The difficulty lies in deciding which to use.

If populations were like rectangles we would probably compare them by length and breadth. These parameters give a good description of size and their relationship provides some idea of shape. We could have compared these 'populations' by other parameters—by the length of the diagonal for instance—but neither that parameter nor any other improves on the description provided by length and breadth.

As with rectangles, so with populations; the usefulness of a set of parameters depends on:

1. ease of estimation;
2. extent to which they collectively describe the significant properties of a population;
3. ability to extrapolate beyond the data from which they were calculated;
4. directness of their relationship to population processes; and
5. their generality—the extent to which they apply to all kinds of populations, not just to some specific populations.

A choice of parameters is a compromise between these attributes. The main parameters used here are:

1. survival by age,
2. fecundity by age,
3. frequency distribution of ages,
4. sex ratio, and
5. numbers or density.

A set of correlate statistics can be calculated from estimates of those parameters:

a. rate of birth,
b. rate of death, and
c. rate of increase.

Other parameters can be used when one or other of those given above cannot be estimated. Fisheries biologists estimate a parameter called 'recruitment', the number of fish added to the catchable stock each year. By all criteria but one this parameter is deficient. It confounds fecundity and juvenile survival, it relates to size rather than to age, and it varies according to fishing methods. But the one point in its favour is rather important; for practical reasons it is often the only estimate by which a fisheries biologist can track fecundity. Our choice of parameters is as often forced by rain, mountains, forests and seas as it is by algebra and logic.

Chapter 3

Age

The most useful parameters are those specific for age. Fecundity, for instance, could be measured as number of offspring produced per female per year, but since the value of this parameter changes with the proportion of young non-breeders in the population, it tells us almost as much (or as little) about the age distribution as it does about reproduction. Fecundity is measured more precisely as the production of young over each parental age interval.

The age of a man can be expressed in months, years, or decades. The choice does not matter much because the human physiology is not strongly geared to intervals of time other than those of 24 hours and 28 days. Time flows for man. The time perceived by most other vertebrates is essentially periodic. All facets of an animal's life—nutrition, hormone balance, behaviour, pelage, growth, reproduction, and chances of survival—are tied closely to the cycle of seasons. Few species escape the domination of seasonal time, the few exceptions being mainly tropical and desertic forms. The year, which is a convenient arbitrary unit of age for man, is a real unit of age for other vertebrates.

If we investigate the effect of age on fecundity and mortality, we must estimate these parameters for a group of animals whose members were born at the same season (although not necessarily in the same year), and the age interval must be set at one year. Otherwise differences between age classes will relate more to season than to age as such. Years are not used as arbitrary units in the following chapters but as natural units of age.

A study of the dynamics of a population depends heavily on an ability to age individuals. Fish are aged by counting annuli on the scales and within otoliths and fin rays, by counting vertebrae and by dividing size distributions into age classes. Reptiles are usually aged by size, the growth curve being constructed either from growth increments recorded in the field from individuals of unknown age or from the growth of individuals captive since hatching. Counts of osteological growth zones have also been used. Since birds are difficult to age by morphological criteria beyond one or two years of age, ornithologists divide birds into juveniles and adults or into young, medium and old age classes. They limit close ageing to those individual banded as fledglings. Mammals can be aged by a variety of techniques, one at least of which is usually applicable to any species being studied. Clues to age are provided by tooth eruption, tooth wear, size frequency classes, body size, degree of epiphyseal fusion, lens weight, annual growth rings on claws and horns and in teeth and bone, and the number of placental or ovarian scars carried by females. Rounsefell and Everhart (1953) described the techniques of ageing fish and Taber (1969) has done the same for birds and mammals. The few gaps in coverage left by these reviews are filled by the publications of Peabody (1961) on osteological growth zones, Low and Cowan (1963) and Mitchell (1967) on growth annuli in teeth, Fabens (1965) on the fitting of a curve to the growth of individuals of unknown age, and Cassie (1954, 1962) and Taylor (1965) on the dissection of a size frequency distribution into its component age classes.

Techniques of ageing need not, therefore, be described here but some of the pitfalls of ageing rate a mention.

Indices of age can be classified as:

    A. Individual marks (e.g. bands and tags placed on young animals at recorded dates)

    B. Morphological indices

        a. Characters that change continuously with age (e.g. lens weight, tooth wear)

        b. Characters that change in quantal jumps

            1. Annual quanta (e.g. growth rings on horns, scales and teeth)

            2. Non-annual quanta (e.g. plumage phases and tooth eruption).

All criteria of age, other than those based on individual marks, are subject to error. Some are worse than others. Indices that change by annual quanta give the most accurate estimates, particularly for adult animals, but they are not foolproof. Morphological characters that change continuously with age, and quantal characters not tied to season, have a variability that automatically results in some ageing errors. Tooth eruption is a good example because it is used frequently to age mammals. Table 3.1 shows the variation in ages at which permanent incisors erupt in Himalayan thar, *Hemitragus jemlahicus* (a mountain goat), and cattle. The cattle were kept in controlled conditions whereas the thar

Table 3.1.   Mean age and variability of eruption of permanent incisors in cattle and thar (Wiener and Donald 1955, Caughley 1965)

Incisor	Cattle		Himalayan thar	
	Mean age (months)	Age interval in which 95% of eruptions occur (months)	Mean age (months)	Age interval in which 95% of eruptions occur (months)
1st	25·1	21·3–28·9	17	12–22
2nd	31·8	27·6–36·0	26	23–29
3rd	37·8	31·6–44·0	42	31–53
4th (C)	45·0	38·0–52·0	64	45–85

ranged freely. Coefficient of variation for ages at eruption is about 8 per cent for cattle and 12 per cent for thar.

The age intervals over which 95 per cent of eruptions occur overlap for thar only between the third and fourth incisors. Yet ageing errors will occur at any point in this sequence of eruption. Suppose a thar were found with a second permanent incisor erupted but with the third socket still occupied by a milk tooth. The best estimate of age is the mid-point between the mean ages of eruption of these two teeth—34 months. Depending on the time of the year, this specimen would be classified either as a two-year-old or a three-year-old. But the actual age might be anything between 23 months (lower 95 per cent limit of eruption for the second incisor) and 53 months (upper 95 per cent limit for the third incisor). The animal might be a two, three or four-year-old.

Continuous measurements produce similar problems. M. J. W. Douglas and I investigated tooth wear as an index of age for thar. We found that from three years of age the regression of age on logged base-crown height of the first molar was linear. Consequently this measurement showed promise as an index of age. However, further study revealed that the standard deviation of ages around the regression had a value of 1·7 years. Hence only 22 per cent of aged specimens would be placed in the correct year class whereas 40 per cent would be shared by the year classes either side of it. The assigned ages of a further 22 per cent would be in error by two years, 10 per cent by three years and 6 per cent by four years. At this stage we decided to abandon the method. Tooth wear was measured rather than assessed visually in that exercise. Ryel, Fay and van Etten (1961) showed that observers differ considerably in their visual assessment of tooth wear, thereby adding to the variance accruing from natural variation a further variance resulting from errors of assessment.

Most ecologists agree that errors in ageing are inevitable. The point in dispute is whether the errors matter much. Robinette et al. (1957) put the optimistic view: 'The writers recognize the errors possible in ageing individual deer. However, it is felt that the prime value of ageing comes from determination of age composition of hunter-killed deer. In this respect, errors should be compensating provided the samples are of adequate size...'. In contrast, Caughley (1967a) argued that errors in ageing do not compensate. The effect of errors can be demonstrated on a model population of only two age classes containing 100 and 10 individuals respectively. If 10 per cent of each class were misidentified by age and placed in the other, the apparent frequencies become 91 and 19. The ratios 100/10 and 91/19 differ enough to distort population statistics calculated from the age distribution. No reduction in bias results from dividing the age distribution by finer intervals of age.

Accurate ageing is important. When an index of precise age is unavailable, as is usually the case, the researcher is forced back to a less dependable technique. But he then needs to know how often his approximate index leads to error, thereby gaining some idea of the confidence he can place in his conclusions. A regression of age on a morphological character (e.g. size, tooth wear, number of ovulation scars) is a dangerous toy unless accompanied by the standard deviation of ages around the regression.

Chapter 4

Abundance

Abundance can be measured in three ways: as the number of animals in a population, as the number of animals per unit of area (absolute density), and as the density of one population relative to that of another (relative density). Although an estimate of population size might seem intrinsically superior to estimates of absolute or relative density, it is often not particularly useful. The number of fish in a pond tells us something but the size of the Saskatchewan moose population has scant biological meaning. The first estimate reveals something about a real population whose members constitute a biological unit. The second population is a unit only in an administrative sense. Population size can only be assigned a biological meaning when the population is circumscribed; but when the boundaries are unknown, or when no distinct boundaries exist, or when the boundary exists only in the minds of men, density rather than size provides the biologically real measure of abundance.

Estimates of abundance have no intrinsic value and they should never be considered ends in themselves. Many biological problems (e.g. various questions on genetics, zoogeography, behaviour, and population management) require no estimate of abundance. Other problems, particularly those linked with utilization of habitat, rate of increase, dispersal, and the reaction of a population to management treatments, can often be solved with estimates of relative density. Only a small number of problems (e.g. sustained-yield harvesting, stochasticized genetics, and that class of studies relating numbers or density on one hand to behaviour, reproduction, survival, emigration and immigration on the other) demand estimates of a population's size or its absolute density.

The majority of ecological problems can be tackled with the help of indices of density, absolute estimates of density being unnecessary luxuries. A density index as used in this sense is any measurable correlative of density. Ideally its trend is linear on absolute density, but non-linear indices are sometimes sufficient. Catch per unit effort, faeces per square metre, density of fish eggs, density of territorial male birds, and number of deer carcases found per unit time or unit area, are examples of indices that reveal something about a population's density. A density index is useful only in comparisons. When one population is compared with another, or the density of a single population is compared between years, density indices are often sufficient measures of abundance. Estimates of absolute density have been used in many studies where density indices would have provided equally enlightening information.

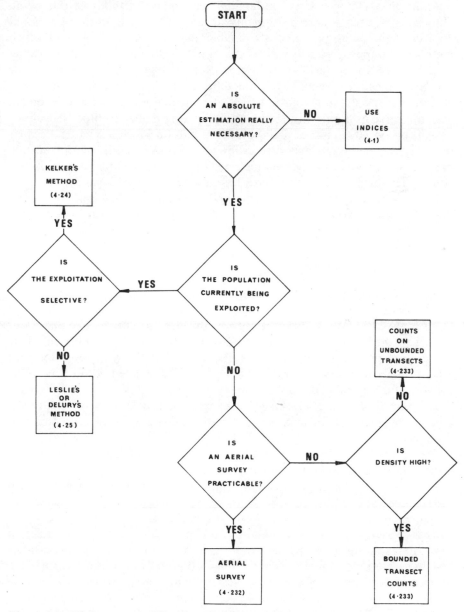

Figure 4.1.   The sequence of decisions by which a census method (other than a mark–recapture technique) is chosen.

A problem defined in terms of absolute density can usually be redefined in such a way that estimates of relative density will provide a solution. If such a design is possible the saving in time, money and stomach ulcers is usually considerable, and the precision of the result is often higher than one based on estimates of absolute density.

Although a biologist must be familiar with many methods of estimating abundance he need not carry the details around in his head. Rather, he should appreciate the range of methods available and have a rough idea of which methods are appropriate to which situations. This chapter and that dealing with mark-recapture methods (Chapter 10) have two purposes. Firstly, they can be skimmed to give an idea of the different methods by which abundance is measured and the situations to which each method is appropriate. Secondly, they can serve as a reference to the details of a particular method, the survey design appropriate to its use, and the assumptions on which it is based. Because mark-release methods are used to estimate statistics of birth, death, and rate of increase, in addition to population size, it is convenient to hold over discussion of these methods until the reader has tackled Chapters 7, 8 and 9.

Figure 4.1 is an aid to choosing a method of estimating abundance by techniques other than mark-recapture. It is a flow diagram of the sequence of decisions leading to the selection of an appropriate method.

4.1 ESTIMATES OF RELATIVE DENSITY (INDICES)

Figure 4.2 diagrams several possible relationships between absolute density and a density index.

Aa is the simplest—a linear regression through the origin. The density of whales might be expected to bear this relationship to the number of whale spouts seen per cruising hour. Ab is also linear but the regression does not pass through the origin. The density of birds is often related in this way to the number of territorial males per unit area.

Diagram B of Figure 4.2 is a negative regression of density on index. The density of a herbivore is sometimes so related to the average height of a preferred food plant.

C is a non-linear positive regression. The density of mice is related in this way to the percentage of traps catching mice per night.

D represents an ambiguous relationship between density and index, each index value implying two possible densities. In some circumstances the density of mammals is related to average body size in this manner.

These various density-index relationships must be interpreted in different ways. Example Aa, which passes through the origin, allows the simplest interpretation. The index doubles when the density doubles, or stated more generally, an increase or decrease of index by a factor of x implies an increase or decrease of density by a factor of x. In contrast, the regression of density on index in examples Ab, B and C, allows only a ranking of density by index except in the rare cases where the y-intercept is known. D does not, of itself, provide even

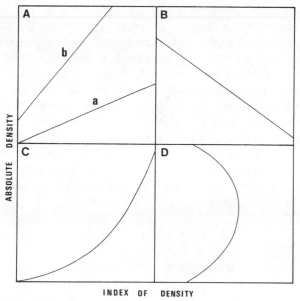

Figure 4.2. Various forms taken by the regression of absolute density on an index of density: (A) linear ascending, (B) linear descending, (C) non-linear, and (D) non-linear recurvate.

a ranking of densities. Quite obviously an index with an Aa relationship to density (linear through the origin) is the most useful. A type C index (curvilinear through the origin) is also useful in that it can often be transformed to make density linear on the index.

In most cases a particular index is chosen because its relationship to density is known to be approximately linear and because there are strong theoretical arguments in favour of the regression passing through the origin. Seldom is the slope of the regression known, and absolute density cannot therefore be estimated directly from the index, but this is unimportant if the question at issue is framed in relative terms. The question: What proportional reduction in the density of deer accrues from doubling hunting pressure? can be answered with type Aa indices of density. For a less specific question such as: Is density influenced by the availability of food (or water or cover)? indices of type Ab, B and C are usually adequate.

4.1.1 Linear density–index relationships

When the regression of density on index is linear the relationship can be summarized as

$$\text{Density} = a + b \text{ Index,}$$

where a is the density when the index is zero and b is the increase of density associated with a one-unit increase of index. In using an index we must know whether b is positive or negative, a judgement usually requiring little more than common sense, and depending on the problem to be solved we may need to know whether a is zero. The latter usually becomes obvious from the answers to the following questions: If there are no animals can the index still be measured? (if so, a is negative) and, When animals are present in low numbers can the index be zero? (if so, a is positive). The second question is not the same as asking whether the index would sometimes be undetectable when density is low. That is a question about sampling variation, not about the functional relationship between density and index. Most indices are direct manifestations of the activity of animals; a low density provides a low index and a zero density provides a zero index, and hence $a = 0$.

Counts of animals

The most commonly used density indices are counts of animals. For instance:

1. number of migrating birds flying across the moon per hour,
2. number of antelopes seen per mile of road,
3. number of deer seen per hour's walk,
4. number of elephants seen at a waterhole per night,
5. number of basking crocodiles counted per mile of river, and
6. number of caribou seen per minute from a plane flying 150 m above the ground.

These examples share three characteristics: we can reasonably expect that the regression of absolute density on index is approximately linear, that it passes through the origin and that its slope is positive. However, assumptions can at the same time be both reasonable and wrong. The regression of deer density on the number seen per hour's walk is often not linear (Thane Riney, personal communication) and there are situations in which each of the indices listed above would not be related linearly to density. If a strictly linear index is needed, the relationship between index and density must be investigated closely before a particular index is accepted as adequate.

Although counts of animals are often used as indices of density they have several disadvantages, the most annoying being that their accuracy depends on strict standardization of the conditions under which they were measured. One does not, for instance, compare the relative densities of two gazelle populations by making road counts through the range of the first at midday and through that of the second at dusk. Nor does one use counts to compare the densities of two populations of lizards, one of which lives in long grass and the other in short grass. Counts of animals also vary according to the ability of the counter. Not only does a skilled observer see more of the animals within his range of vision but he also causes less disturbance. Fewer animals move away from his line of travel before he has a chance to see them. The act of observing is itself an influence on how many animals are counted—physics holds no patent on Heisenberg's uncertainty principle.

Counts of animal signs

Counts of animal signs, rather than counts of animals, often provide a more accurate index of density. All animals leave a record of their presence in the form of such clues as tracks, faeces, cast antlers, disturbed or grazed vegetation, kills (if they are predators), pelagic eggs, abandoned nests, and so on. An index based on animals signs has these advantages over counts of animals: its accuracy is less dependent on the skill of the observer, its measurement is easier to standardize between observers, it is affected less by viewing conditions, and the act of observing does not influence what is observed. Its disadvantages, in comparison with an index of animals counted, lie in its less direct relationship to actual density and in the time lag between a sign being laid down and being observed. Signs provide an index of mean density over a period of time rather than an index of current density.

The density of animal signs is sampled in exactly the same way as the density of animals. Appropriate designs and analyses are discussed in Section 4.2.3.

Catch per unit effort

Catch per unit effort provides a useful index of density when the catching does not greatly reduce the size of the population. When the catches are large relative to the size of the population, absolute catch per unit of time is a better index.

Catch per unit effort should never be used as an index of density unless the population is being exploited. It requires more labour, causes more disturbance, and in general is less accurate than most alternative indices. But if the study population is being investigated at the same time as it is being hunted or fished, or if the investigator must take specimens for purposes other than estimating density, this index can be calculated with little extra work. It is particularly useful as a cross-check on other estimates of relative density.

The use of catch per unit effort assumes that:

1. Conditions of catching are standardized. There is little point in comparing the catch of fish per boat per day in two areas when one catch was taken during a storm and the other during good weather.

2. Catching efficiency is standardized. Deer shot per man per day will not provide a comparison of density between two areas when the first is hunted by professionals and the second by sportsmen.

3. Gear must be standardized. Obviously no comparison is possible between fish netted per day and fish hooked per day, but more subtle variation in

gear must also be watched for. The number of small mammals captured per trap-night depends not only on density but on the kind of trap and the kind of bait. The number of large mammals shot per man-day seems to show little association with the calibre or make of rifle in use but can vary greatly according to whether the rifle is fitted with telescopic or open sights.

4. The catching of one animal should not interfere with the catching of another. Fisheries biologists speak of 'gear saturation', the clogging of a net or trap, but this is only one aspect of the problem. The presence of one animal in a trap may inhibit others joining it (Kennedy 1951) or may attract others (Ricker 1958). The number of large mammals a hunter shoots, relative to the number he sees, depends during an open season on average group size. When groups of two or three animals are encounted he will usually shoot all animals in the group, but as group size increases the proportion of animals shot per group decreases. This is an effect of 'time saturation' rather than gear saturation. The hunter can shoot only so many animals before the survivors are out of range. Single-catch traps provide a special case of gear saturation. Catch per trap per night can never be greater than 1 and the regression of density on catch per trap-night is therefore curved. However, below an index of about 0·2 the regression is almost linear, competition for traps having little effect on the number caught. Above 0·2, catch per trap-night becomes an increasingly inaccurate measure of relative density and the index must be rectified by the method given in the section 'Frequency measurements' below.

5. Animals must not learn to court or avoid capture. Capture-shyness and its obverse, capture-proneness, the twin demons that dog most studies of small mammals, are the results of unintended education. Capturing methods liable to generate these effects should not be used if rate of capture is employed as an index of density. However, if other methods of capture are impracticable, the trapping period should be limited so that the animals have little chance of learning how to avoid or to seek capture.

When the five conditions listed above are met, the regression of absolute density on catch per unit effort is linear through the origin (Ricker 1940).

Examples of the skilful use of this index are provided by the work of Leslie and Davis (1939) on small mammals, Gulland (1955) on fish, Riney (1956) and Batcheler and Logan (1963) on alpine ungulates, and Child, Smith and Richter (1970) on African game. Dasmann and Dasmann (1963) used an index of this type to show that deer in California passed through a surge of density between 1936 and 1943. They graphed the ratio of deer killed to hunting licences issued each year between 1927 and 1960. Figure 4.3 presents their two sets of data—number of deer tags sold and the number of deer reported as killed—and the index of density (hunter success ratio) that was calculated from them. The ratio could rightly be subjected to a frequency–density transformation (Section 'Frequency measurements' below) but since it generally holds to a value below 0·2 the untransformed ratio is probably a sufficiently accurate index of abundance.

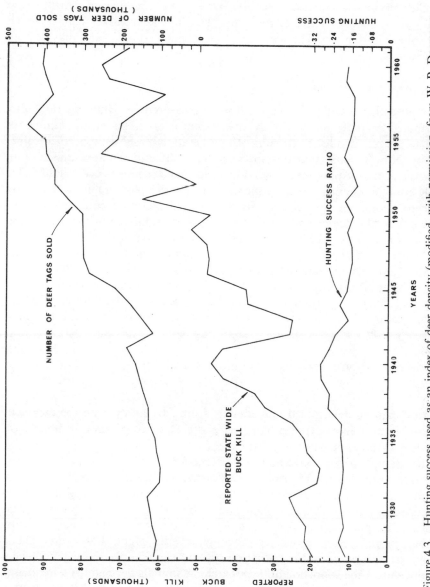

Figure 4.3. Hunting success used as an index of deer density (modified, with permission, from W. P. Dasmann and R. F. Dasmann, *Calif. Department Fish and Game*, **49**, 4–15, 1963).

4.1.2 Non-linear density–index relationships

When the regression of absolute density on index is curved a doubling of index no longer implies a doubling of density. The indices can be used only to rank densities. This major disadvantage can be eliminated either by calculating the curvilinear regression to give an equation predicting density from index, or by rectifying the relationship. The first alternative requires much additional information. The second is usually much simpler and may require no further data.

Frequency measurements

Suppose the density of a population is measured by counting randomly distributed animals on randomly distributed plots. The mean number of animals per plot divided by the area of a plot is an unbiased estimate of absolute density. Now suppose that this species is very shy and the appearance of an observer at a plot puts many individuals to flight. The observer could judge easily enough whether or not animals were on the plot but he would have difficulty estimating how many. His data would therefore be in the form of 'presence or absence per plot' and it would be summarized as the proportion of plots containing one or more animals. This statistic is a 'frequency' as opposed to a 'density' and its relationship to absolute density is non-linear.

When animals are randomly distributed, mean frequency per plot, f, can readily be transformed to mean density per plot, $\bar{x}$. The proportion of plots containing 0, 1, 2, ... animals is given by the Poisson distribution of which the first term is $e^{-\bar{x}}$. The proportion of plots containing no animals is therefore

$$1 - f = e^{-\bar{x}}$$

and the proportion containing one or more animals is

$$f = 1 - e^{-\bar{x}}.$$

Hence $\bar{x}$ can be determined immediately from f by looking up the exponent of $1 - f$ in a table of exponentials (Appendix 1), or by reading it directly off Figure 4.4 if an approximation is sufficient.

The example given above may not be encountered often in the field but it illustrates the density-frequency transformation which is applicable to a broad range of ecological data. For instance, the number of small mammals caught in single-catch traps is often used as an index of density. Each time an animal is caught one trap is put out of action and the number of active traps is thereby reduced progressively throughout the night. Leslie and Davis (1939) recognized that this situation provides a simple density-frequency relationship. The proportion of traps catching animals per night is a frequency related to the number of animals that would have been caught per trap if the traps were capable of multiple captures. The latter variable is a 'catch density' which can easily be calculated from the 'catch frequency'.

Suppose that the relative densities of rats in area A and area B were measured

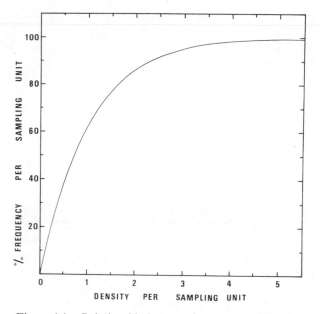

Figure 4.4.   Relationship between frequency and density.

by setting 100 single-catch traps for one night in each area, and that 40 rats were caught in the first area and 80 in the second. These catch frequencies are converted to catch densities as follows :

	A	B
Frequency of capture per trap (f) :	0·4	0·8
Proportion of traps catching none ($1 - f$) :	0·6	0·2
Estimated density of catches per trap, $\bar{x}$, when $(1 - f) = e^{-\bar{x}}$:	0·5	1·6

The transformed indices indicate that the density of rats in area A is not twice that of area A, as might be deduced from the untransformed frequencies of capture, but more than three times as high.

Presence or absence of animals per plot, and frequency of capture in traps, are obvious candidates for a frequency-density transformation. Several other indices of abundance are less obviously in this category but must be transformed in this way to rectify their relationship to density. Baits eaten per night, disturbance of cotton threads stretched across tracks, proportion of food plants browsed, proportion of cover slips laid out on the ground that were broken by trampling, proportion of muddied water holes, and hunter success when the bag limit is one kill, are all frequencies that must be transformed to linear indices of density.

The use of the frequency-density transformation is well illustrated by two studies of the marsupial *Trichosurus vulpecula*. Batcheler, Darwin and Pracy (1967) used it to convert trapping frequencies to density estimates, and Bamford (1970) transformed the frequency with which baits were disturbed to indices linear on animal density.

When a frequency is below about 0·2, frequency is almost linear on density and no transformation is necessary. A frequency of 0·05 plots in which animals are present implies a density of 0·051 animals per plot. When frequency rises to 0·2 the equivalent density is 0·22 animals per plot. Above 0·2 frequency and density diverge at an accelerating rate and an untransformed frequency is no longer a useful density index.

The validity of the frequency-density transformation rests on the assumption that all individuals have an equal probability of occurring on a plot at the time of sampling, or in the case of trapping, that all individuals are equally catchable. These assumptions are usually violated in practice because of clumped distributions in the first case and varying catchability in the second. Both deviations lead to underestimates of relative density, the bias downward increasing with density. Consequently, a comparison of density indices from two areas will usually provide an underestimate of the difference. If there must be a bias, this is the way it should be. Underestimates lead to fewer unfortunate management decisions than do overestimates.

Gerrard and Chiang (1970) devised a modified frequency-density transformation that can cope with non-random distributions. Instead of expressing frequency as number of plots containing one or more animals, they counted the number of plots containing j or more, when j is any whole number. They showed that for any level of contagion there is an optimal j which can be determined from a pilot trial. Their method necessitates much initial work to estimate a transformation, but thereafter the calculation of density is quick and accurate, in fact often more accurate than complete counts of animals per plot.

Nearest-neighbour distances

A sample of distances from a random point (or a randomly chosen individual) to its nearest neighbour provides a non-linear index of density. Absolute density can be estimated from these distances if the individuals are distributed randomly. When w is the *square* of a nearest-neighbour distance and $\bar{w}$ is the mean of a sample of n squared distances, the mean number of animals on an area of one square unit of measurement is estimated by

$$N = \frac{n-1}{\pi \bar{w} n}$$

If 50 distances were measured in metres to give a mean squared distance of 220, density is

$$N = \frac{50 - 1}{3 \cdot 1416 \times 220 \times 50}$$

$$= 0 \cdot 0014 \text{ animals per m}^2 \text{ or } 1400 \text{ per km}^2.$$

The equation returns an unbiased estimate of density only when the animals are randomly distributed, and since they never are the formula might seem to be only an academic curiosity. However it does have a practical use. It transforms nearest-neighbour distances to linear indices of density. Provided that the order of contagion does not change with density the regression of absolute density on the transformed index is not far off a straight line.

Nearest-neighbour distances do not, of themselves, reveal whether animals are distributed at random (Pielou 1969 : 114). Some kinds of contagious distributions produce a spread of distances indistinguishable from the pattern generated by a random distribution. Batcheler and Bell (1970) provided a modified transformation that returns an estimate of absolute density even when the distribution is moderately contagious. Since it treats random and regular distributions as special cases of contagion it does not require a prior evaluation of the pattern of distribution.

Density classes

It often happens that only a rough index of density is required. An administrator responsible for the management of quail over an entire state might wish to know whether a uniform code of hunting regulations would suffice, or whether hunting seasons should differ in length in different parts of the state. Amongst other things he would need to know whether or not the density of quail was about the same throughout the state. For this question small differences in density are operationally equivalent to no differences. If one person in each county reported whether quail were 'absent', 'rare', 'common' or 'abundant', the administrator would have information on density adequate for his purposes.

These categories are called density classes, and estimates of abundance in this form are sufficiently accurate to answer a surprisingly broad range of questions. The disadvantages of density classes are immediately obvious:

1. one man's 'common' class may be another man's 'abundant' class,
2. absolute density is usually not linear on a scale of density classes, and
3. being subjective, their accuracy depends on the skill of the observer.

The major problem with density classes is actually a non-problem. People tend to regard them as more accurate than they are in fact, and they are often used on problems whose solution demands more refined estimates. When the intrinsic lack of sensitivity of this kind of estimate is recognized, and their use is limited to studies requiring only a broad evaluation of density, they can be very useful.

The accuracy of density classes is enhanced when the criteria of 'rare', 'common' and 'abundant' are defined. If 'common' is defined as 'between 20 and 50 animals seen in a day's walk', the estimate is repeatable. Standards differ between regions and drift with time. When a nineteenth century explorer wrote that kangaroos were 'common' he meant that one or two were seen each day. Today, if a stockman reports that kangaroos are 'common' he means that he sees between 50 and 100 during a day's ride.

The human mind, for some reason, tends to classify density on a logarithmic rather than an arithmetic scale. If an observer's undefined 'common' density averages about three times his 'rare' density, then his 'abundant' density is liable to average three times the 'common' density and hence nine times the 'rare' density. When density classes are numbered 1, 2, 3 and so on, the regression of absolute density on these is straighter when they are converted to antilogs.

'Presence or absence' is a commonly used index of density. It might appear to be an objective dichotomy, but it is subject to the same difficulties as more extended classifications. These estimates are always biased downwards because although a 'presence' is usually unambiguous, 'absence' means that animals are either present or absent. It is well to remember that 'absent' means that none were seen, not that none were there. An 'absence' record based on a search from a vehicle is not the same as one resulting from a walk through the area. The accuracy of a 'presence or absence' survey is closely related to the time spent observing.

Sometimes this does not matter. For some problems a low density has the same implications as absence. A biologist mapping the distribution of a pest species with a view to controlling it where it causes damage is no more interested in an area of low density than he is in one where the species is absent. Low density and absence are equivalents to him. But if the project were aimed at conserving a species threatened with extinction, 'rare' and 'absent' have very different implications and the two classes must be differentiated. In the first case a cursory survey provides the information on which a management decision is made; in the second a more detailed survey is called for.

Animals per group

The mean size of social groups of gregarious animals tends to increase as density increases. The reason for this is not fully understood, but it can safely be accepted as an empirical observation.

If the composition of social groups were fluid, groups splitting, coalescing and exchanging individuals, a regression of group size on density would be imposed by the correlation between absolute density and the frequency with which groups meet. Caughley (1964) suggested such a mechanism to account for the regression of density on group size for kangaroos. However, the non-random composition of groups of ungulates and other social animals indicates that this mechanism is not determining group size for these species, and the explanation is probably simplistic even for kangaroos. Irrespective of what the mechanism might be, group size usually provides a workable index of density. Christie and Andrews (1964) showed that it could be used to define density classes of thar, and it may be equally useful for many other social species.

Mean group size can usually be calculated very accurately and with high repeatability. If only for this reason it is always worth considering as a density index. It does have disadvantages: the relationship of index to density changes with season and varies between habitats. It is useful only for comparing the density of populations at the same time of the year in areas of similar topography and vegetation. The regression of density on group size is seldom linear and it usually cuts the vertical axis below the origin. Hence the index ranks densities but does not reveal the proportional differences between them.

4.2 ESTIMATES OF ABSOLUTE DENSITY

4.2.1 Total counts

A count of the total number of animals inhabiting a study area is sometimes feasible when the area is small and the animals are conspicuous. Censuses of this kind have been attempted for alpine ungulates (Douglas 1967), mammals on the Serengeti plain (Talbot and Stewart 1964), wildfowl on lakes (many studies), and so on. A total count is possible only when the animals are relatively sedentary or when the survey is run over a period short enough to preclude significant movement. Otherwise some animals will be counted twice and others missed completely.

4.2.2 Guesses

When a man is thoroughly familiar with an area and its fauna he may be able to make a realistic guess at the number of animals living there. If an accurate estimate is not required a guess may serve as an estimate of density. The numerous and obvious disadvantages of a guess should not be allowed to obscure the advantages. Sometimes an informed guess is closer to the truth than an objective estimate made by mark-recapture, change-of-composition, or other indirect methods. Whether or not it is more accurate, a guess at density or numbers provides a useful cross-check on an objective estimate. If the indirect method returns a total considerably higher or lower than an informed guess, a healthy scepticism in the accuracy of the objective estimate is entirely in order, and the reality of the assumptions on which it is based should be investigated further.

At the same time, guesses should be recognized for what they are. They are not 'ocular estimates' or 'survey evaluations' and their accuracy is not enhanced by giving them these fancy names. Guesses are guesses. We must not fall into the trap of thinking that because we are thoroughly familiar with an area and with a species living there our guess will necessarily come close to the true total. Even highly skilled observers usually underestimate.

An example of the extent to which informed guesses can be in error is reported by Andersen (1953). The rangers managing a population of roe deer in Denmark estimated that about 70 lived in a forested area of 3·4 km². Since they spent much of their time in this forest there was good reason to suspect that their estimates would be reasonably accurate. But when this population was shot out to make way for another strain of deer the total was found to exceed 213 animals. The guesses of the rangers were out by a factor of three.

Guesses should be backed by information allowing some appraisal of their probable accuracy: how long was the man in the area; was his estimate based on counts of tracks, signs, calls or animals, or a combination of these; and how experienced was he? A simple check on the accuracy of a guess is the consistency of guesses made independently by several people. In 1924 the size of the Kaibab deer herd was variously estimated as 100,000, 70,000, 60,000, 50,000 and 30,000 (Rasmussen 1941, Caughley 1970a). The variability within this set of guesses indicates that little reliance can be placed on any one of them.

4.2.3 Sampled counts

Although total counts appeal intuitively as an accurate method of estimating density, in fact sampled counts have several advantages over them:

1. Sampled counts require less work,
2. They greatly reduce the chance of counting some individuals more than once and missing others completely,
3. They need not, in contrast to total counts, be completed over a short period of time, and
4. the population is less disturbed by the census.

Most statistical models of sampling assume that animals distributed at random are counted on plots located at random. In practice, animals are not distributed at random but tend to have a clumped distribution. Likewise, only in the most favourable circumstances is it possible to lay out plots at random. Such sampling is more often an idealistic aim than an attainable goal. The overriding practical principle of sampling is that the distribution of plots, while it may or may not be exactly random, must in no circumstances be biased.

Density is sampled by dividing the area under survey into sampling units and counting animals on a preselected proportion of these. The mean density per unit sampled is taken as an estimate of the mean density on sampled and unsampled units combined. The confidence that can be placed in this estimate is calculated from the variation in density between the sampled units.

Two influences on the accuracy of the estimate outweigh all others. Firstly, if we divide an area containing 100 animals into ten units and sample only two of these, it is possible that, by chance, neither unit will contain animals. The resultant grossly inaccurate estimate of zero density would probably have been avoided if four units were sampled. Secondly, suppose that the 100 animals constitute a school of fish in a lake. At any one time the school is likely to occupy only one sampling unit. When two units only are sampled, the density per sampling unit will be estimated either as zero or as 50, implying respectively that either no fish or 500 fish live in the lake. These examples illustrate two principles of sampling: the greater the number of units sampled the greater the accuracy of the estimate, and the more clumped are the animals the more sampling units needed to give a reasonably accurate estimate of density.

Stratification

So far we have assumed that animals are evenly distributed throughout the surveyed area, even though the pattern of distribution, viewed in detail, will be clumped. More often the area comprises a number of plant associations, some of which contain animals at high density whereas others harbour few individuals. In these circumstances the most accurate estimate of mean density is obtained by dividing the area into zones within which density is homogeneous and combining the densities measured separately in each zone. The process is called stratification and the zones are called strata.

Stratification has these advantages:

1. Because densities are estimated separately for each stratum these densities can be compared. Since the strata usually represent different habitats this information is biologically important.

2. Sampling effort is allocated most efficiently when the intensity of sampling increases with density. There is little point in sampling intensively a stratum containing few animals. Even if the estimated density in this zone is double the true density, the error will have little effect on the estimate of the average density over the whole area.

3. The precision of an estimate of density is inversely related to the variability of density throughout the area. Stratification divides an area of heterogeneous density into a number of strata within which density is relatively homogeneous. By this process the precision of the estimate is increased because it is now a function of variability of samples within strata, not of variability across the entire area. Siniff and Skoog (1964) demonstrated that their stratified sampling of caribou returned an estimate with double the precision that would have resulted from simple random sampling.

Stratification presupposes a knowledge of the distribution of animals. Ideally, the area is divided into strata before the survey begins and the proportion of units sampled per stratum is allocated according to density. Formally, the estimate of total density is most precise (i.e. has the smallest standard error) when the intensity of sampling is proportional to the standard deviation of animals per sampling unit within the stratum (Tschuprow 1923, Neyman 1934). A pilot survey to obtain rough estimations of these standard deviations will often be justifiable economically. Certainly, if the survey is one of a series, sampling effort per stratum should be allocated according to the results of the previous survey. If a pilot survey is not possible, and if the area has not been surveyed previously, the relative density in each stratum should be guessed on a scale of one to ten. Sampling intensity is then allocated proportional to this index because, for most species, the standard deviation of animals per sampling unit tends to vary linearly with mean density per sampling unit.

The placement of strata boundaries is not critical. Large errors in placement will increase the standard error but will not bias the estimate of density. When density is clearly heterogeneous, even clumsy stratification is preferable to no stratification.

The analyses appropriate to stratified sampling are presented in the section following. Although that section deals specifically with the problems of aerial survey, the analyses are quite general and are fully appropriate to any sampling design aimed at estimating density by direct counting.

Aerial surveys

Census from the air is an important technique of wildlife management. The problems associated with it are amply covered by the papers of Petrides (1953), Fuller (1953), Banfield *et al.* (1955), Gilbert and Grieb (1957), Stewart, Geis and Evans (1958), Erickson and Siniff (1963), Bergerud (1963), Siniff and Skoog (1964), Goddard (1967), Watson *et al.* (1969), Caughley (1974) and LeResche and Rausch (1974). The East African Agricultural and Forestry Journal (Vol. 34, 1969) published an excellent symposium on problems and techniques of aerial census. In this issue Jolly (1969) described experimental designs and analyses appropriate to aerial census and the present treatment closely follows his approach.

Figure 4.5 diagrams four survey designs for an area that can be divided into two density strata. The first (A) represents stratified random sampling with sampling units of equal area. In design (B) the area is divided into units of unequal area, the probability that a particular unit is selected being proportional to its size. The sampling units of design (C) are strips of fixed width and direction but placed randomly. Design (D) is the same as (C) except that the strips are regularly spaced. The analyses appropriate to these four designs, and the problems associated with them, will be examined in turn.

Equal-sized units. A map of the area to be surveyed is gridded and the rows and columns of the grid assigned numbers. Pairs of random numbers are taken from a table, the first representing the row and the second the column. The grid square indicated by the two numbers is marked for sampling. This process is repeated without replacement until the number of marked sampling units in each stratum reaches the total previously decided on. For the example in Figure 4.5 (A), 50 per cent of the high-density stratum is sampled as against 25 per cent of the low-density stratum. The selected units are flown and the number of animals seen on each is recorded.

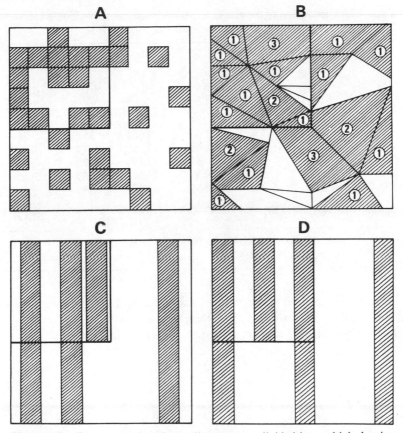

Figure 4.5. Four strategies of sampling an area divided into a high density stratum (the left upper square in each map) and a low density stratum: (A) by sampling equal-sized units at random, (B) by sampling at random unequal-sized units whose probability of selection is proportional to their areas, (C) by selecting at random transects oriented in one direction, and (D) by sampling systematically spaced transects. In each case sampling intensity on the high-density stratum is double that on the low-density stratum.

Notation is taken mainly from Cochran (1963) who outlines the theory and methods of stratification. The subscript h denotes the stratum and i the sampling unit within the stratum. The following symbols all refer to stratum h.

N_h = total number of sampling units
n_h = number of units sampled
y_{hi} = number of animals on the ith unit
$\bar{y}_h$ = mean number of animals per unit sampled
s_{yh}^2 = variance of numbers between sampled units
$Y_h = N_h \bar{y}_h$ = estimated total number of animals in the stratum.

The calculations will be demonstrated with Siniff and Skoog's (1964) data

Table 4.1. Calculation of the size of a caribou population by stratified random sampling. Data from Siniff and Skoog (1964)

(1)	(2)	(3)	(4)	(5)	(6)	(7)
Stratum	N_h	n_h	$\dfrac{N_h(N_h - n_h)}{n_h}$	s^2_{yh}	$\bar{y}_h$	$Y_h = N_h\bar{y}_h$
A	400	98	1,233	5,575	24·1	9,640
B	30	10	60	4,064	25·6	768
C	61	37	40	347,556	267·6	16,324
D	18	6	36	22,798	179·0	3,222
E	70	39	56	123,578	293·7	20,559
F	120	21	566	9,795	33·2	3,984
Totals	699	211				$Y = 54,497$

Sum of products of columns 4 and 5 = 34,305,121
$\text{S.E.}_Y = \sqrt{34,305,121} = 5,857$

from a survey of caribou. They defined six strata and sampled within each a number of 4 ml^2 units. Table 4.1 gives their figures.

Total numbers in each stratum are estimated as

$$Y_h = N_h\bar{y}_h = 9640 \text{ in } A$$
$$768 \text{ in } B$$
$$\cdot$$
$$\cdot$$
$$\cdot$$
$$3984 \text{ in } F.$$

The number of animals in the total area is therefore the sum of strata totals:

$$Y = \sum Y_h = 54,497$$

which differs slightly from the estimate of Siniff and Skoog (1964) who used another method of calculation.

A standard error of Y is calculated from columns 4 and 5 of the table:

$$\text{S.E.}_Y = \sqrt{\sum_h \frac{N_h(N_h - n_h)}{n_h} s^2_{yh}}$$

where s^2_{yh} is calculated for each stratum as

$$s^2_{yh} = \frac{1}{n_h - 1}\left(\sum y^2_{hi} - \frac{(\sum y_{hi})^2}{n_h}\right).$$

Hence $\text{S.E.}_Y = \sqrt{34,305,121} = 5857$ and 95 per cent confidence limits are assigned as $\pm 2 \text{ S.E.}_Y = \pm 11,714$.

Unequal-sized units. The previous method is used only when all sampling units are the same size. More commonly we sample on strip transects of constant width (usually 100 metres either side of the plane) but of varying lengths. The precision of the estimate is highest when flight lines are laid out randomly as in Figure 4.5 (C), but regularly spaced transects as in (D) of the same figure have practical advantages that cannot be ignored.

Designating the area of a sampled unit as z_i and total area under survey as Z,

$$Y = \frac{\sum_h N_h \bar{y}_{h_i}}{\sum_h N_h \bar{z}_{h_i}} Z$$

where Y is again an estimate of the total number of animals in all strata. To calculate the standard error of Y we first define

$$R = Y/Z$$

and

$$s_{hzy} = \frac{1}{n_h - 1} \left[\sum_i z_{hi} y_{hi} - \frac{\left(\sum_i z_{hi} \right) \left(\sum_i y_{hi} \right)}{n_h} \right]$$

to give

$$\text{S.E.}_Y = \sqrt{\sum_h \frac{N_h (N_h - n_h)}{n_h} (s_{yh}^2 - 2Rs_{hzy} + R^2 s_{zh}^2)}$$

with s_{yh}^2 being the variance of numbers per sampled unit in the hth stratum and s_{zh}^2 the variance of the areas of the sampled units in the same stratum. The calculation of S.E.$_Y$ is an elaboration of the method used in the equal-area example, and it can be estimated in much the same way by tabulating the data as in Table 4.1, to which are added columns of s_{zh}^2 and s_{hzy}.

Equal or unequal-sized units selected with probability proportional to area. In mountainous country linear transects are impracticable, as are sampling units of equal area; the area must be divided into units that can be flown freely.

Sampling units with straight-line boundaries are drawn on a map. They can be of any shape, but three and four-sided figures simplify calculation of areas. Units are again selected from each stratum by pairs of random numbers, but they are used in this case as map coordinates to define points, not areas. Selection of random pairs proceeds until each stratum contains the required number of points. Those units containing one or more points are marked for sampling. A unit is counted in the calculations as many times as the number of random points that fell within it, although it need be surveyed only once. Figure 4.3(B) shows a set of sampling units chosen in this way, the number in each unit indicating the number of points that fell within it.

This method of selection ensures that the chance of selecting a particular unit is proportional to its size, a property that has several statistical advantages and which leads to simplified calculations.

The data are first converted to densities, D_i, for each unit by $D_i = y_i/z_i$. The number of animals in the hth stratum is then estimated as

$$Y_h = z_h \bar{D}_h$$

where z_h is the total area of the stratum and $\bar{D}_h$ is the mean density over the units sampled within the stratum. The total number of animals in the whole area is estimated by summing the total from each stratum: $Y = \sum Y_h$. Its standard error is calculated as

$$\text{S.E.}_Y = \sqrt{\sum_h \frac{z_h^2}{n_h} s_{Dh}^2}$$

where

$$s_{Dh}^2 = \frac{1}{n_h - 1} \sum_i D_{hi}^2 - \frac{(\sum D_{hi})^2}{n_h}.$$

Systematic versus random sampling. The analyses given in the previous section are each based on the assumption that sampling units are randomly located. Although the estimation of numbers is most precise when this requirement is fulfilled, in so doing practical problems are encountered which may often result in a loss of accuracy overwhelming the modest gain accruing from immaculate survey design. These problems are:

1. Random location of sampling units places a considerable strain on the navigator. The unit boundaries are difficult to determine when landmarks are few; many mistakes are made. The system works best when the biologist knows the area well, but then he must divide his time between counting animals and instructing the pilot. If he flies the plane himself his observing efficiency drops and the count suffers in consequence.

2. Planes are noisy and wild animals are timid. The combination results in movement of animals immediately after an overflight. Hence sampling units must be well spaced to avoid movement from one unit to another during the survey. Some units will be close to or contiguous with others if they are located at random, thereby resulting in double counting of animals that move between adjoining units.

3. Random location results in the lowest coverage of sampled area per hour of flying.

Although systematic sampling is less sound theoretically, it is not hampered by the practical problems listed above. Evenly spaced flight lines greatly ease the strain of navigation. So long as the pilot fixes his starting point accurately (lines are best started from a winding road or river that provides navigational clues), he can hold direction by compass while checking for drift as he passes prominent landmarks. Flight lines are usually spaced at intervals of two or three km. Animals moved off a line by disturbance of the plane seldom reach the next line before the plane returns. Systematic sampling results in a minimum of dead time between counts. Although this is not necessarily an advantage, since observers need a rest every 15 minutes or so, it is an important consideration when aiming at the best estimate of density from a fixed allocation of money or time.

Many books on sampling warn that the estimate of confidence limits from sampled counts is valid only when samples are taken at random. And so it is, within the restricted definition of 'valid' favoured by statisticians. A statistical conclusion is valid only when the data on which it is based were collected according to the axioms underlying the appropriate statistical model. Axioms such as the normal distribution of variables, independence of observations and randomness of sampling underly most statistical tests, and on most occasions that these tests are used the axioms are violated, sometimes slightly, sometimes grievously.

The validity or otherwise of a statistical procedure is not the most important consideration. We search for a robust test rather than a valid test, one that will give an answer close enough to the truth even when the data do not fully fit the axioms. The question we ask of a statistic procedure is not whether its use is valid but whether it is sufficiently robust to embrace the deviations from underlying axioms that we know lurk within the data. Sometimes we must decide that the data are so different from the kind of data envisaged in the model that we cannot use the test. Sometimes we can transform the data to something more like what the test is supposed to deal with; and sometimes we just use the test on the raw and invalid data, but drop the acceptable significance level from 5 per cent to 1 per cent to be on the safe side. Mathematical statistics is an exact science, but the use of statistical tests on biological data is most inexact.

A confidence limit calculated from non-random samples may be invalid but it is seldom much different from a confidence limit calculated from random samples. So long as the sitting of systematic transects is not biased with respect to what lies on the ground, the axioms of the statistical model are not grossly violated. Of course, if the topography of the surveyed area forms a systematic pattern and the flight lines are layed out along the grain of the country, the estimate of animal numbers may be highly inaccurate. Or if we counted ducks on lakes but surveyed only lakes known to hold large populations, the estimate of total numbers would be very wrong. These survey methods are invalid in the commonly accepted sense of the word, and it is this sense that should be kept in mind when deciding whether or not to use a particular sampling design.

Techniques of aerial survey. For the reasons outlined previously, populations are usually censused from the air along flight lines spaced at regular intervals. The animals are counted on a 100 m strip on both sides of a high-wing plane flown 75 m or so above the ground. The boundaries of this strip are demarcated for an observer by two streamers attached to the wind strut. In the gap between them he sees a 100 m strip of ground when the plane is at survey altitude. Although the streamers can be positioned by geometric calculation, an empirical positioning is less chancy. A permanent marker is placed on the strut close to the fuselage (the inner marker) and beyond it, at regular intervals up the strut, a set of numbered temporary markers are attached. The plane is flown at survey altitude alongside a 100 m strip marked out on the ground, the observer guiding the pilot onto a course that positions the inner marker on the inner boundary of the strip. The correct position of the outer marker is then noted as the point on the strut coinciding with the observer's line of sight to the strip's outer boundary. Streamers are attached to the inner and outer markers to provide linear sighting boundaries (Figure 4.6).

Systematic sampling is not possible in mountainous terrain. Random sampling with the probability of selecting a unit being proportional to its area is the most appropriate design. The survey goes more smoothly if one observer navigates while the other counts. Although many pilots are capable and willing to help with counting this practice should not be encouraged. Piloting in these conditions requires undivided attention and any lapse in the pilot's concentration can well lead to truncated results.

Since observers must not take their eyes off the transect during an overflight most record their counts onto tape. Having had a couple of unfortunate experiences with faulty tape recorders I prefer a notebook. The transect is divided into segments of two minutes flying, each separated from the next by an obligatory pause of seven seconds. The area flown over in those seven seconds is excluded from the transect and the number seen on the last segment can be

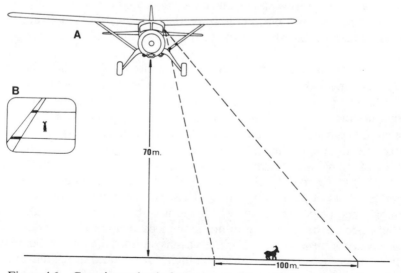

Figure 4.6. Counting animals from the air (A), and visual delineation of transect boundaries by strut markers and streamers (B).

jotted down during the dead time. That system has additional advantages. It allows an observer to change the focus of his eyes, to straighten the kink in his neck, and to check with the other observer that they agree on the serial number of the segment. It also acts as an antidote against aerial survey hypnosis, the symptoms of which can easily be recognized as an inability to focus, a tendency to day-dream, and sporadic loss of memory such that the observer cannot recall whether or not he has recorded his last observation.

A camera is essential if the animals are aggregated. An accurate count of a large group is impossible during the short time it is in sight. Experienced observers can often make a close estimate by counting very quickly in fives or tens, but a photograph allows an accurate estimate of numbers. The camera must be positioned at the observer's normal line of sight, and the photograph must include the boundary streamers so that animals outside the survey strip are not included in the count. Watson (1969) discussed the combinations of lens, altitude, film and lighting that give the best results.

Correction for undersampling. Few observers, including myself, can believe that they miss animals on a transect while flying at 75 m in good light. They

Table 4.2. Efficiency of aerial censusing

Species	Percentage counted	Conditions	Control[a]	Reference
Black rhino	29	Plain and scattered scrub	good	Goddard (1967)
Indian rhino	56	Jungle and elephant grass	poor	Caughley and Mishra (unpublished)
African game	< 88	Plain and scattered timber	poor	Lamprey (1964)
Man	65	Plain (fixed wing)	good ⎰	Watson, Jolly and
	75	Plain (helicopter)	good ⎱	Graham (1969)
Simulated game	75	Table model	exact	Watson, Freeman and Jolly (1969)
Moose	77	Snow and scattered timber	poor	Edwards (1954)
Wapiti	< 64	Grass and scattered timber	poor	Buechner, Buss and Bryan (1951)
Mule deer	< 43	Light timber	fair	Gilbert and Grieb (1957)

[a]The accuracy of the aerial survey is measured by how close the total came to an estimate made by a different, and hopefully more accurate, method. The accuracy of these control totals is also variable, the rating given in this column being my subjective assessment of reliability based on the quoted authors' description of their method. A rating of 'poor' indicates that their control estimate is probably too low, and that the estimated percentage counted from the air is therefore too high.

can accept this intellectually because it has been proven every time it has been tested, but it hardly seems possible. I once checked the accuracy of an aerial survey by sitting behind the regular observer and looking out the same side of the plane. We were counting African mammals on a grassed plain. At the end of each line we compared results and found that, in general, our tallies were similar although mine were a little lower, the result to be expected since he was more familiar with the species. As a final check we flew a transect for which we each called an observation as we made it. The difference was marked. He saw animals that I missed and I saw animals that he missed. Even giraffes, which are very conspicuous, were sometimes missed by one or other of us. This sobering exercise is recommended to any observer who thinks he sees 95 per cent of animals on a transect.

Table 4.2 gives several estimates of the efficiency of aerial survey, showing that numbers are invariably underestimated. The error is usually large enough to invalidate aerial census totals, although they are useful, not withstanding, as indices of abundance or as estimates of minimum numbers. But if they are to be used as estimates of absolute density they must be corrected. Caughley and Goddard (1972) showed from computer simulations and censuses of black rhinoceros that true totals could be estimated from the mean and variance of repeated censuses made at two levels of survey efficiency. The method is demonstrated on Gilbert and Grieb's (1957) counts of mule deer (Table 4.3). Deer were counted from the air on an area of approximately 20 km^2 in Colorado. During the first series of five surveys the scattered snow cover made for difficult

Table 4.3. Counts of mule deer from the ground and from the air (Gilbert and Grieb 1957)

Time and conditions		Aerial count replicates	Ground count
1st census		246	
		229	
		220	
(patches of snow)		238	
		255	
	$\bar{x} =$	237·6	695
	$s^2 =$	189·3	
		339	
2nd census		312	
		345	
(snow cover		353	
almost complete)		342	
	$\bar{x} =$	338·2	690
	$s^2 =$	241·7	

observing. By the time of the second set of surveys a month later, the area was blanketed with snow and observing conditions were excellent. Hence the records provide two sets of surveys run at different levels of survey efficiency.

The model has the form

$$\bar{x} = s^2 \frac{1}{k} + \bar{x}^2 \frac{1}{N}$$

where N is the true population size

$\bar{x}$ is the mean number counted per survey at one level of survey efficiency
s^2 is the variance of these counts, and
k is a constant describing the extent to which mean sightability varied between surveys at one level of survey efficiency.

Since we have two unknowns, k and N, a solving is possible only when two estimates each of $\bar{x}$ and s^2 are available. For the example

$$\begin{cases} 237 \cdot 6 = 189 \cdot 3 \frac{1}{k} + 237 \cdot 6^2 \frac{1}{N} \\[2ex] 338 \cdot 2 = 241 \cdot 7 \frac{1}{k} + 338 \cdot 2^2 \frac{1}{N} \cdot \end{cases}$$

Multiplying the first by $241 \cdot 7$ and the second by $189 \cdot 3$ gives

$$\begin{cases} 57,428 = 45,754 \frac{1}{k} + 13,644,874 \frac{1}{N} \\[2ex] 64,021 = 45,754 \frac{1}{k} + 21,651,990 \frac{1}{N} \end{cases}$$

from which, by subtraction of the first from the second

$$6593 = 8,007,116 \frac{1}{N}$$

and

$$N = 1214 \text{ deer in the area.}$$

From this result we can theorize that only about 24 per cent of the deer were counted during any one aerial survey and that the ground counts tallied only about 58 per cent of the deer in the area at the time.

In this example the extent of snow cover fortuitously provided a difference in survey efficiency between the two sets of surveys. More commonly we would create the difference ourselves, either by flying one set of surveys at 100 m and the other at 200 m or by increasing the number of observers on the second set of surveys.

Because this method operates on total counts it is not useful for estimating true density over an area that can only be sampled. An alternative method is then used (Caughley 1974) in which true density is estimated as the y-intercept of the regression of observed density per sampling unit on speed (S), height

(H) and transect width (T). For red kangaroos in open country, for example, the predictive equation has been estimated (Caughley *et al.* 1976) as

$$\text{Per cent counted} = 100 - 0 \cdot 1718\,S - 0 \cdot 0123\,H - 0 \cdot 007\,H^2 - 0 \cdot 0982\,T,$$

from which a correction factor can be calculated for any combination of survey variables. On a transect of 100 m per observer viewed from 75 m above the ground at 160 kph, an observer sees only 58 per cent of kangaroos on his transect. The count from each sampling unit must therefore be multiplied by a correction factor of $1/0 \cdot 58 = 1 \cdot 72$ before further analysis.

Ground surveys

Where animals can be counted from the air they can also be counted from the ground. Survey design and analyses are similar, with the minor difference that the width of a transect must usually be reduced in ground counting to allow for reduced visibility. For each survey design appropriate to aerial survey there is an analogous design appropriate to ground counting. The technique of counting from the air along fixed flight lines, for instance, has a direct parallel in counting from a vehicle fitted with a range finder, and the data are analysed in the same way.

The major difference between counts from the air and counts from the ground lies in the greater disturbance caused by the second method. Aerial surveys certainly cause disturbance but the animals react to the plane after it has passed, not before it arrives.

Such is not the case with ground surveys. If an observer zig-zags through a sampling unit counting timid and mobile animals, by the time he has covered half the area few animals will remain on the other half. Ground surveys must therefore be designed around the behaviour of the animals. If they are highly mobile, sampling units should be long and narrow. Sometimes accurate ground surveys are impossible: when animals live in cover with a mean visibility of 10 m but flush, on average, at a range of 20 m, no survey design will allow an estimate of density. Indirect methods must be used.

Eberhardt (1968) classified the conditions influencing counts of animals:

Detection depends primarily on

- searching by the observer
  - readily visible animals
  - limited visibility or cryptic animals
- conspicuous response of an animal to observer's approach
  - fixed flushing radius
  - probability of flushing depends on distance from the observer

The survey design must be shaped by these conditions. Most problems of ground counting can be tackled by variations on two strategies: by counting on a transect of fixed width, the width being determined by the animal's behaviour and sightability, or by counting all animals seen from a line of march.

Transects of fixed width. The first strategy leads to a more simple analysis and it is particularly appropriate when densities are high. The distance from the line of march to the transect boundary paralleling it on either side is set at the distance within which all animals present will be seen. It may be a metre when frogs are counted, 10 m when large mammals are counted in heavy forest, or 500 m when animals of this size are censused in grassland. The decision on whether an animal is 'in' or 'out' is critical; a range finder is necessary when the transect is wider than 100 m. If the censused species is undisturbed by human presence an animal should not be tallied until the observer reaches the point closest to it on his line of march. From this position he is best able to estimate the 'right-angle distance', and hence to decide whether the animal is within the transect. But individuals belonging to timid species, or those that are seen as they flush, must be judged immediately as inside or outside the transect. Unless it is obvious, a compass angle and a distance from the observer is needed to calculate whether the animal should be tallied.

Experimental design and analysis for fixed-width transects is the same as that given in Section above 'Aerial surveys' for unequal-sized sampling units. The area is stratified on a map, both by density of animals and by visibility. Within a stratum transect width is determined by sightability, and the appropriate number of transects is determined by the size of the stratum, the transect width, and animal density. The last can be estimated roughly by a small pilot survey.

Transects of indefinite width. Eberhardt (1968) argued that since the precision of a density estimate is proportional to the square root of the number of animals counted during a survey, it is highly desirable that all animals seen should be recorded. Data of this kind are analysed differently from counts on transects of fixed width. The following additional information is needed:

1. the right-angle distance of each animal from the line of march (the fixed-width transect method requires only an 'in' or 'out' decision),
2. form of the curve relating the probability of seeing an animal to the distance it stands from the line, and
3. the slope of this curve.

Since each of these estimates is subject to error, the decision on whether to use transects of fixed or indefinite width must balance the simplicity and reduced experimental error of the first against the added precision accruing from the greater number of animals counted by use of the second. In general, the second design is more appropriate at lower densities.

Experimental design and analysis will be illustrated by an imaginary example. Suppose we are estimating the density of a species of bird living in grassland.

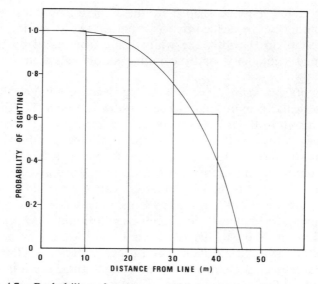

Figure 4.7. Probability of seeing an animal at varying distances from a line of march.

Right-angle distances are recorded for all birds seen either side of a 10 km line of march. These records can be summarized as a histogram (Figure 4.7) constructed by classifying the observations into 10 m intervals and dividing the frequency of each interval by that of the first to form a probability distribution. We assume that the number of birds recorded in the first interval (within 10 m of the line) represents a complete count. The probability of seeing a bird in this distance is therefore set at unity, and beyond, the probabilities fall away with distance from the line.

Eberhardt (1968) suggested a flexible curve to fit this kind of data:

$$P_x = 1 - \left(\frac{x}{W}\right)^k,$$

where P_x is the probability of seeing an animal located x metres from the line and W is the maximum distance at which an animal can be seen. The constant k describes the shape of the curve. When $k = 1$ the regression is linear, convex when $k > 1$ and concave when $k < 1$.

Table 4.4 outlines the steps in fitting a curve to the probabilities in Figure 4.7. The constants k and W are estimated from the regression of log $(1 - P_x)$ on log x for all intervals other than the first. The regression constants a and b estimate respectively $- k$ log W and k. Table 4.4 and Figure 4.7 show that the values of k and W estimated in this way generate a curve closely tracking the observed decline of sightings with distance.

Having fitted this curve and checked that it mimics the data, we can now

Table 4.4. Fitting a curve to the probability of seeing a bird at varying distances from a line of march

Distance (m) from line	Mid distance x	Birds seen	Probability of sighting P_x	Used for regression		Fitted probability P'_x
				$\log x$	$\log(1 - P_x)$	
0–10	5	50	1·00			1·00
–20	15	49	0·98	1·1761	– 1·6990	0·98
–30	25	43	0·86	1·3979	– 0·8539	0·88
–40	35	31	0·62	1·5441	– 0·4202	0·61
–50	45	5	0·10	1·6532	– 0·0458	0·07
		$n = 178$				

Regression: $\log (1 - P_x) = k \log x - k \log W$
$$= 3 \cdot 444 \log x - 5 \cdot 723$$
$$k = 3 \cdot 444; \; W = 45 \cdot 92 \text{ m}$$

Hence $P'_x = 1 - \left\{\dfrac{x}{45 \cdot 92}\right\}^{3 \cdot 444}$

Mean distance $= \bar{x} = \dfrac{(5 \times 50) + (15 \times 49) + \ldots + (45 \times 5)}{178} = 18 \cdot 93 \text{ m}$

Number of sightings $= n = 178$ birds
Length of line $= L = 10{,}000 \text{ m}$

estimate density. By Eberhardt's (1968) model the density per unit area, D, is solved as

$$D = \frac{n(k + 1)^2}{4L\bar{x}k(k + 2)}$$

where $n =$ the number of animals seen $= 178$
$\bar{x} =$ the mean of their right-angle distances from the line $= 18 \cdot 93$ m
$L =$ length of the line $= 10{,}000$ m.
For the example,

$$D = \frac{178 \times 4 \cdot 444^2}{4 \times 10{,}000 \times 18 \cdot 93 \times 3 \cdot 444 \times 5 \cdot 444}$$

$$= 0 \cdot 000248 \text{ birds per m}^2$$

$$= 248 \text{ per km}^2.$$

It is informative to compare this estimate with that from a transect of fixed width. If the first distance interval is used to define the width of the strip, since we assume that all birds were seen on this strip, density is estimated as

$$D = \frac{\text{number seen on the transect}}{\text{length} \times \text{width}}$$

$$= 50/(10{,}000 \times 20)$$
$$= 0{\cdot}000250 \text{ birds per m}^2$$
$$= 250 \text{ per km}^2.$$

Note that the transect width is not 10 m, the first distance interval, but 20 m, since birds are counted up to 10 m either side of the line of march. This estimate is about the same as by the previous analysis, but since it is based on a count of only 50 birds, in contrast to the 178 records from the transect of indefinite width, its precision is only half ($\sqrt{50}/\sqrt{178} = 0{\cdot}53$) that of the previous estimate.

A model proposed by Höglund, Nilsson and Stålfelt (1967) and discussed by Eberhardt (1968) assumes that the probability of sighting falls away exponentially with distance. No bounding distance need therefore be assumed and the estimate of density simplifies to

$$D = \frac{n}{2L\bar{x}}.$$

Eberhardt gave theoretical reasons for suspecting that this model would be appropriate when detection depends on visibility rather than on the behaviour of the animals. It may prove to be the best estimate for mortality surveys and for ground surveys of African game. Its applicability can be tested by graphing the log of numbers seen per interval of distance against the mid-points of these intervals. When the regression is approximately linear the exponential model yields a satisfactory estimate of density.

4.2.4 Selective additions and removals (Kelker's method)

If a population is classifiable into two or more classes—say according to colour phase, sex, or age—its size can be estimated after a known number of animals has been added to or removed from one class. The method necessitates an initial survey to determine the ratio of one class to the other, a reduction or addition of individuals heavily selective of one class, and a second survey to determine the new ratio. Kelker (1940, 1944) introduced this technique as a means of estimating the number of deer in a population hunted only for males. Subsequently Petrides (1949), Chapman (1954, 1955), Lander (1962), Chapman and Murphy (1965) and Rupp (1966) elaborated and generalized the method.

Designating the two classes as x and y, and

N_1 = population size at the first survey,
N_2 = population size at the second survey,
p_1 = proportion of x-individuals in the N_1 population,
p_2 = proportion of x-individuals in the N_2 population,
C_x = number of x-individuals added to or removed from the population between surveys (additions are positive, removals negative),

C_y = number of y-individuals added to or removed from the population between surveys (additions are positive, removals negative), and
$C = C_x - C_y$,

the population's size at the time of the first survey is

$$N_1 = \frac{C_x - p_2 C}{p_2 - p_1}.$$

Suppose that we are studying a population of lizards and take a sample to check on reproductive condition of females. As a by-product of this sampling we can also estimate the size of the population. Firstly, we estimate the proportion of females in the population; let us say it is $p_1 = 0.64$. The sample is then collected: 50 females ($C_x = -50$) and ten males ($C_y = -10$). A second survey immediately afterwards estimates the proportion of females in the population as $p_2 = 0.51$. The population size is therefore

$$N_1 = \frac{(-50) - 0.51(-60)}{0.51 - 0.64}$$

$$= \frac{-19.4}{-0.13}$$

$$= 149 \text{ lizards of both sexes.}$$

The method allows great flexibility of experimental design since C_x and C_y may both be positive, both negative, one positive and one negative, or one either positive or negative with the other zero. Kelker's method is often used by wildlife biologists to estimate the size of a game population prior to hunting that removes mostly males. Sex ratio is estimated before and after the hunting season and the number killed during the season is established at checking stations. Since the method copes equally well with additions (Rupp 1966) it can also be used in such problems as estimating the size of a pheasant population into which game-farm cocks are released.

Chapman (1955), Ricker (1958) and Rupp (1966) suggested a further modification whereby x and y are two distinct species rather than two classes within one species. Suppose we wished to estimate the numbers of a game species that lived in the same area as a protected species. C_x is then the number removed from the game population and $C_y = 0$ the number removed from the protected species. When one species is used in this way as a control on the reduction of a second, we are not particularly interested in the combined total of the two populations but in the size, N_x, of only one population.

Here

$$N_{x1} = \frac{p_1(C_x - p_2 C)}{p_2 - p_1}.$$

The 'two species' method is particularly appropriate when a species of frog or lizard is studied in an area containing one or more additional species that are closely related taxonomically and ecologically to the species of interest. The population size of the species under study can thereby by estimated without manipulating the population or disturbing its members.

Kelker's method depends on these assumptions, that

1. the two classes are equally available at each survey,
2. there is no natural mortality between surveys,
3. there is no recruitment or immigration between surveys, and
4. all removals and additions are recorded.

Chapman and Murphy (1965) have extended the method to allow for natural mortality. They provide equations to estimate the rate of exploitation both when animals are removed over a short period of time and when exploitation continues throughout the year.

4.2.5 Non-selective additions and removals

Leslie's method

In the absence of recruitment and natural mortality, catch per unit of effort plotted against previous cumulative catch provides a straight line cutting the x-axis at the population size prior to harvesting (Leslie and Davis 1939). When all catches are obtained with equal effort, as they might be in a designed

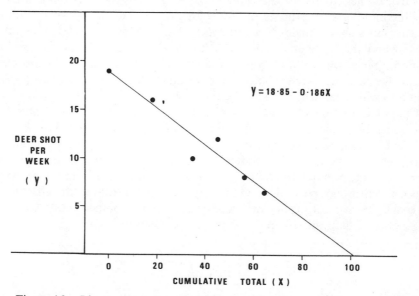

Figure 4.8. Linear regression of yield per week against cumulative yield for use in Leslie's method of estimating population size.

experiment (in contrast to catches of a commercial fishery), the y-axis can be graduated in catch units (e.g. tons of fish, number of animals) rather than in units of catch per effort.

Figure 4.8 is a typical plot of catch per unit effort against cumulative previous catch. The slope of this regression (with sign reversed) is the "catchability", the proportion of the population taken with one unit of effort. In this example the slope of $b = -0.186$ indicates that, on average, two riflemen each week shot 18·6 per cent of the deer present in their hunting area at the beginning of the week. The x-intercept estimates the population size, N_1, before they began hunting:

$$N_1 = -a/b = -18.85/-0.186 = 101 \text{ deer.}$$

Zippin (1956) pointed out that the least-squares method of fitting the regression biases the estimate slightly and he provided a weighting factor to correct for this effect. However, Ricker (1958) considered that since the scatter of points in determined more by day-to-day variation in catchability than by statistical influences, the increase in accuracy obtained by weighting is seldom worth the effort.

Leslie's method depends on three assumptions. As modified from Moran (1951) they are:

1. probability of being caught is constant for all animals on each catching occasion,
2. the population is not so dense that the catching of one individual interferes with the catching of another, and
3. no births, deaths, immigration or emigration occur during the experiment.

Few biologists would accept even one of these assumptions as realistic, but the experiment can be designed to minimize the difference between the model's assumptions and the behaviour of the animals. The probity of the first assumption is indicated by the linearity of the regression. If the trend of the points is obviously not linear the method should be abandoned. The second assumption can be made true by careful experimental design. If, for instance, single-catch traps are used, enough traps must be set such that no more than 20 per cent are sprung during a trapping interval.

Assumption 3 comes close to reality when the experiment is run over a short interval of time. The number of catching intervals should be reduced to a minimum and the catch in each of these should be as large as possible. Zippin (1956) and Seber and Le Cren (1967) gave analyses appropriate to this design. They showed that when only two catches, C_1 and C_2, are taken with the same effort, the population prior to the first catch is estimated by

$$N_1 = C_1^2/(C_1 - C_2)$$

and the catchability, p, by

$$p = (C_1 - C_2)/C_1$$

If it is assumed that p as estimated above holds for a second population also, that population's size can be estimated from a single catch (Seber and Le Cren 1967) by

$$N_1 = C_1/p.$$

In practice this would be a chancy procedure because the size of a single catch is influenced so much by the weather, but careful matching of conditions might result in an adequate estimate.

Leslie's method works equally well when, instead of removing animals from the population, the catch is marked and released again. On subsequent catching occasions the marked animals are ignored and only those unmarked are recorded as 'caught'. But this method, actually a variant of mark-recapture, is relatively inefficient. A Petersen estimate (Section 10.3) yields a more precise indication of population size for the same expenditure of effort. The variance of estimates made by Leslie's method are large relative to those of alternative models. The method should be used only when alternatives are impracticable, or when a control campaign or cropping scheme can be exploited to provide estimates as by-products of these management activities.

DeLury's method

DeLury (1947) showed that logged catch per unit effort is linear on effort expended prior to a given catching occasion. The slope, b, of this regression is related to catchability, p, by

$$1 - p = \text{antilog}_{10} b,$$

and the initial size of the population is estimated by

$$N_1 = \frac{C}{1 - (1 - p)^n}$$

where C is the total catch and n is the number of units of effort.

Table 4.5. Estimating population size by DeLury's method

Week	Previous effort (wks)	Catch	$\log_{10}$ catch
1	0	25	1·40
2	1	26	1·42
3	2	15	1·18
4	3	13	1·11
5	4	12	1·08
6	5	13	1·11
7	6	5	0·70

$b = -0.1007$; $1 - p = \text{antilog}_{10} b = 0.793$; $p = 0.207$
$C = 25 + 26 + \ldots + 5 = 109$
$N_1 = C/1 - (1 - p)^n = 109/(1 - 0.793^7) = 136$ fish

The close relationship between this method and Leslie's is obvious, and in fact they can be derived from the same equation (Ricker 1958). Hence, the constraints on the use of this method are the same as those listed in the previous section.

DeLury's method will be demonstrated on data collected by Harte (1932) and analysed by Ricker (1949, 1958). Whitefish were caught in Shakespeare Island Lake, Ontario, over seven successive weeks. Size of nets, their locations and the time of setting were standardized, and hence a week's fishing can be considered as one unit of effort. Table 4.5 gives the catch figures and their analysis.

Index-manipulation-index

Population size can be estimated from a linear index of density measured before and after a known number of animals are added to or removed from the population. The same analysis is used in both cases except that additions enter the calculations as positive numbers whereas removals are negative.

Riney (1957) studied a population of red deer, *Cervus elaphus*, living in 30 km^2 of forest in New Zealand. He estimated an index of density by counting the number of fresh and medium-aged faecal pellet groups on 2160 4-m^2 plots, calculating a mean of 0·17 groups per plot. Two professional hunters were then assigned to the area and they shot 66 deer over the next $2\frac{1}{2}$ months. A further month later the index of density was measured again. It had dropped to 0·07 fresh and medium groups per plot.

Hence we have:

index of density before the hunt . . . . $I_1 = 0\cdot17$
index of density after the hunt . . . . $I_2 = 0\cdot07$
and, number removed during the hunt. . . . . . . $C = -66$.

The pre-hunt population size is estimated as

$$N_1 = \frac{I_1 C}{I_2 - I_1}$$

$$= \frac{0\cdot17 \times (-66)}{0\cdot07 - 0\cdot17}$$

$$= 112 \text{ deer,}$$

and the post-hunt size by

$$N_2 = \frac{I_2 C}{I_2 - I_1}$$

$$= \frac{0\cdot07 \times (-66)}{0\cdot07 - 0\cdot17}$$

$$= 46 \text{ deer.}$$

The precision of the estimate is dependent on the assumption that the population is closed during the course of the experiment. In practice some animals usually die naturally during this interval and estimates of N_1 and N_2 are thereby both biased downward. To minimize the bias, the three steps of the experiment—index measurement, manipulation, index measurement—should be crowded into the shortest possible time.

This method is particularly appropriate when a game population is supplemented by individuals raised at a game farm or hatchery. Since individuals are added, not removed, C is positive. There is, however, a trap here. The additions must have no more influence on the level of the second density index than is warranted by their numbers. Hence when hatchery-reared fish are added to a trout population, catch per unit effort cannot be used as an index of abundance because reared fish are more easily caught than wild fish.

4.2.6 Corrected index

Index and control

Occasionally we find a situation in which the species we are interested in lives in an area containing a second species whose density is known. The density of the study species X can be estimated from the known density of Y if we have indices of abundance of both X and Y.

The method is illustrated by a census of kangaroos in Queensland, Australia. Kangaroos and sheep were counted from systematic transect lines in a timbered paddock of 15 km^2 known to contain 1430 sheep. In all, 24 kangaroos and 160 sheep were counted. Symbolizing the unknown number of kangaroos as N_x, the known number of sheep as N_Y, and the number of kangaroos and sheep counted as n_X and n_Y, the number of kangaroos in the paddock is estimated as

$$N_x = \frac{N_Y n_X}{n_Y}$$

$$= \frac{1430 \times 24}{160}$$

$$= 214 \text{ kangaroos.}$$

The known density of a species can be used to estimate in this way the density of a second species only when their indices of abundance are equivalent. In terms of the example, sheep and kangaroos are assumed to react equally to the presence of an observer, to be equally conspicuous and to be dispersed in a similar way throughout the area under survey. A known density of impala, for instance, could be used to estimate the density of hartebeeste. It could not be used to estimate the density of lions because these are difficult to find during the day, whereas impala and hartebeeste are about equally conspicuous at any distance from an observer.

Calibrated index

When the relationship between a density index and absolute density is known across a broad range of densities, any single value of the index can be converted to an estimate of the true density.

The number of data needed to establish the regression of density on index depends on the complexity of the relationship. When the regression is known to be linear through the origin its slope can be estimated from one measure of density at a known level of index. At least two points are needed when the regression, although linear, misses the origin, and three points are minimal data for estimating the trend of a curvilinear regression.

A simple example is provided by the relationship between density measured by aerial survey and true density. In most cases the regression will be linear through the origin. If the number of animals on one km² is accurately counted as *D*, and the number counted on the same area from the air is *I*, all aerial counts for this habitat type can be converted to true density by multiplying them by D/I. In practice several estimates of this correction factor would be needed to give an accurate conversion. When the regression of density on index is linear but not through the origin, the correction takes the form

$$D = a + bI$$

The constants *a* and *b* are estimated by routine regression analysis of two or more pairs of density–index estimates.

A curvilinear relationship between density and index can be tackled in three ways:

1. When the general form of the regression can be deduced theoretically the index can usually be transformed such that the regression of density on index is rectified. Thereafter a correction factor is calculated as outlined previously.

2. The relationship between density and index may be transformed empirically. At least three pairs of data are necessary, and several more are required if an accurate correction factor is sought. Various transformations, of which log and square root of one or both variables are the most useful, are applied in turn and the resultant regressions are inspected for deviation from linearity. When an acceptably linear trend is found the correction factor is estimated from the constants of the linear regression equation. An advantage of this method lies in the ease with which correction factors are calculated subsequently for other habitats. If the log of density is linear on the square root of index in habitat A, there is a good chance that the same relationship holds also for habitat B, although the slopes and *y*-intercepts of the regressions may differ. Hence the correction factor appropriate to the latter habitat can be calculated with considerably fewer data than were needed to establish the form of the index-density relationship in the first place.

3. Instead of rectifying a non-linear trend, a curve can be fitted directly to the data to provide a regression equation that will convert any index value

to its equivalent density. The polynomial regression is perhaps the most useful here because it generates a wide variety of curves, one of which usually fits the data. This regression takes the general form

$$D = a + b_1 I + b_2 I^2 + b_3 I^3 \ldots,$$

and the terms are added successively (the method is given in most statistical texts) until a satisfactory fit is achieved.

When one variable is predicted from another, that predicted is called the dependent variable and it is graphed, by convention, on the y-axis. Density is the dependent variable in this case, and the correction factor that converts index to density must be calculated from regression of density on index, not from index on density. Certainly, in a biological sense, an index depends on density and not vice versa, but 'dependence' in the statistical sense carries no causal connotations.

A regression calculated by least-squares analysis is unbiased only when the independent (x-axis) variable is fixed. A regression of population size on time, for instance, is an example of a random variable regressed on a fixed variable. An estimate of time has no formal error attached to it. But when estimated absolute density is regressed on estimated index, both variables are random and the estimate of slope is biased downward slightly. When the regression equation is subsequently used to convert indices to densities the resultant values are more likely to be underestimates than overestimates.

Chapter 5

Rate of increase

This chapter is an introduction to 'rate of increase' sufficient to carry the reader over the next three chapters. It will be amplified later, particularly in Chapter 9.

5.1 FINITE AND EXPONENTIAL RATES

The simplest measure of a population's rate of increase is the ratio of numbers in two successive years. This is the finite rate of increase, sometimes termed the growth multiplier (e.g. Rogers 1968), which temporarily will be labelled λ. Thus

$$\lambda = \frac{N_{t+1}}{N_t},$$

N being numbers and t being time. When λ is greater than 1 the population has increased between time t and $t + 1$; when less than 1 the population has declined.

The numbers in a population increasing at a constant rate conform to the sequence:

Year: 0   1   2   3   etc.
numbers: N_0 $N_0 \times \lambda$ $N_0 \times \lambda \times \lambda$ $N_0 \times \lambda \times \lambda \times \lambda$ etc.

which is expressed more concisely as

N_0 $N_0 \lambda$ $N_0 \lambda^2$ $N_0 \lambda^3$ etc.

and generalized as

$$N_t = N_0 \lambda^t.$$

Most papers on vertebrates give the rate of increase as a percentage formed by subtracting 1 from λ and then multiplying by 100. Thus $\lambda = 1 \cdot 2$ would be expressed as 20 per cent. This practice causes confusion when the rate is high. A rate of increase expressed as 170 per cent is difficult to interpret; the statement that the population is multiplied by $2 \cdot 7$ per year allows no ambiguity.

The symbol λ will henceforth be replaced by e^r. It comprises a constant e and a variable exponent r. The constant is the base of natural (= Naperian) logs, taking the value 2·71828. The exponent r is the power to which e is raised such that $e^r = \lambda$; r is the exponential rate of increase. Although the replacement of the simple statistic λ by the more complex e^r may seem barbaric, it leads to simplified algebra and a better appreciation of the nature of a rate of increase.

The finite rate of increase e^r relates numbers at two points of time by

$$N_t = N_0 e^{rt}$$

which can also be written as

$$N_0 = N_t e^{-rt},$$

or

$$\frac{N_t}{N_0} = e^{rt},$$

or

$$\log_e N_t = \log_e N_0 + rt.$$

The last version is particularly useful. Because $\log_e e = 1$, e drops out and the equation takes the general form $y = a + bx$, the formula for a straight line.

Rate of increase can be expressed either as the finite rate e^r or the exponential rate r. The second is more useful for these reasons:

1. It is centred at zero whereas e^r is centred at unity. Hence a rate of increase measured as r has the same value as an equivalent rate of decrease, apart from reversal of sign. If the rate of increase of one population were measured as $e^r = 1\cdot649$, and a second returned $e^r = 0\cdot607$ no direct comparison of the rates reveals that the rate of increase of the first equals the rate of decrease of the second. But when these rates are expressed as $r = 0\cdot5$ and $r = -0\cdot5$ the relationship is obvious.

2. In contrast to e^r, r converts easily from one unit of time to another. When r per year equals x, r per day equals $x/365$.

3. A useful indication of the rate' at which a population grows is provided by the doubling time. It is easily calculated from r by $0\cdot6931/r$. When rate of increase, measured on a yearly basis, is $r = 0\cdot23$, the population doubles every three years. Should $r = -0\cdot23$, the population halves every three years.

For these reasons r is used in preference to e^r. One can be converted easily to the other by

$$\log_e e^r = r.$$

Most books of mathematical tables provide natural logs, and these can also be calculated from logs to the base 10 by

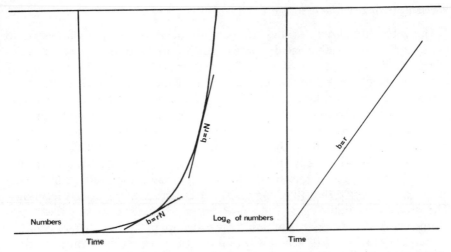

Figure 5.1. The left-hand diagram graphs the size of a population growing at a constant rate. At any point this line has a slope of rN. The same data are graphed in the right-hand diagram as natural logs, the regression having a slope of r.

$$\log_e x = 2 \cdot 3026 \log_{10} x.$$

Appendix 1 gives conversions of e^r to r.

Figure 5.1 shows that the trend of numbers graphed against time is an exponential curve when rate of increase is constant. The slope of the curve changes continuously, but it can be represented at any point by a tangent of slope rN. This simple relationship between slope, exponential rate of increase, and numbers, accrues from the use of natural logs. For this reason logs to the base e rather than to base 10 are used in population analysis. The trend of $\log_e$ numbers on time is linear with a slope of r when rate of increase is constant.

5.2 MEASURES OF INCREASE

Intrinsic rate of increase

The intrinsic rate of increase, symbolized r_m, is the exponential rate at which a population with a stable age distribution (Section 9.5) grows when no resource (e.g. food, space, shelter and water) is in short supply. The rate is determined by the population's genetic constitution interacting with the quality of the environment (Andrewartha and Birch 1954), where quality refers to weather, suitability of food, suitability of nesting sites and so in. The intrinsic rate of increase could equally well be called the 'intrinsic rate of decrease' because r_m is negative for most species in most places. In the Antarctic r_m is positive for emperor penguins. The same species liberated in the tropics would have a negative r_m. It might find there an abundance of food and other resources but it would not be able to cope with the high temperature and humidity.

Survival–fecundity rate of increase

This rate, symbolized r_s, is the exponential rate at which a population would increase if it had a stable age distribution appropriate to its current schedules of age-specific survival and fecundity. Those terms will not be defined rigorously until Chapters 7 and 8, but the general idea is simple enough. A rate of increase depends on the mean fecundity at each age and mean survival at each age. When these hold constant for a period of time the age distribution stabilizes and the rate of increase becomes constant. The survival–fecundity rate of increase is that rate implied by age-specific survival and fecundity at the time of measurement. It usually differs from the actual rate of increase because an age distribution is seldom stable. But the age distribution reflects influences in the recent past rather than the current patterns of survival and fecundity. The survival–fecundity rate of increase, being computed in a way that strips off the effect of the age distribution, reveals the population's current capacity to increase. It also reveals what the actual rate of increase will become if current fecundity and survival remain constant.

Rate r_s differs from r_m in that its estimation is not limited to instances when resources are in superabundance. Hence r_m is a special case of r_s.

Observed rate of increase

A completely general measure of increase is the observed rate of increase, $\bar{r}$. It is the exponential rate at which a population increases over a period of time. It carries no implication that the rate of increase is constant over this period, that the age distribution is constant, or that resources are in superabundance.

These three rates—r_m, r_s, and $\bar{r}$—are mutually dependent and occasionally congruent. A sparse population increasing at rate r_m is also increasing at the rate $r_s = r_m$ and $\bar{r} = r_m$. The rates r_s and r_m are easily confused because they are calculated in the same way from schedules of age-specific fecundity and survival (Chapter 9), the only difference being the circumstance of their measurement. Whereas r_m is the maximum rate at which a population could increase in a specified environment, r_s is the rate at which it could increase at given levels of resources per head.

5.3 DEMOGRAPHIC VIGOUR

Rate of increase reveals much more about a population than the speed with which it grows. It measures a population's general well-being, describing the average reaction of all members of the population to the collective action of all environmental influences. No other statistic summarizes so concisely the demographic vigour of a population.

Both r_m and r_s measure demographic vigour, but in different ways. Birch (1960) suggested r_m as an ecological index of a population's genetic fitness.

In this context r_m measures the innate capacity for increase in a particular environment. Except when density is low, it does not measure the actual rate at which a population increases. Thus r_m is a measure of potential for growth, and the actual rate of growth will usually be reduced below this level by shortages of food, space and shelter. The rate r_m is the appropriate statistic for comparing the fitness of two species in one environment, or for comparing the favourability of two environments occupied by one species.

In contrast, r_s is more general; it weighs up and combines the vigour of each sex and age class of the population under the conditions, including shortage of resources, that a population actually faces. Any change in the environment will result in a rapid change in r_s, whether the environmental change is the addition of further animals to the population, a bout of good weather, an induced change in habitat or the drying of a water hole. The favourability or otherwise of the environmental change is faithfully reflected by the direction and degree of change in r_s. When r_s is viewed in this way it is being employed as a measure of the population's fitness, not in any long-term sense but in the sense of the immediate reaction to the present environment. In this context r_s will be termed 'demographic vigour' to distinguish it from r used simply as a growth constant. The demographic vigour of a population is defined as its level of well-being in terms of fecundity and survival, as summarized by r_s. It tells nothing about how a population will respond to a sudden environmental change, nor whether the population will survive a further ten years, but it does reveal how the population is coping with its current problems. Although demographic vigour means very little in a genetic or evolutionary sense, it is the fitness that is manipulated by management.

Several indices of a population's well-being are in use—fat reserves, body size, sex ratio, age ratio, weight—which are more easily measured than is r_s. They are held to provide indices of a population's 'condition', with 'condition' usually being either undefined or described in vague terms. Such indices are useful so long as we know just what it is that they index. If one population has a higher 'condition' (as measured by an index) than another, how is the difference manifested ecologically? This semantic difficulty is outflanked if 'condition' is equated with demographic vigour. The indices are then useful to the extent that they predict r_s. An index that does not correlate with r_s doubtless measures something, but it does not measure the chances of surviving and reproducing, and therefore does not reveal much about the elemental processes by which a population reacts with its environment.

Potential rate of increase

The idea expressed by 'demographic vigour' can be modified for harvested populations. If the harvesting is carried out skilfully r_s will be stabilized at zero, but the vigour of interest to the game manager is the value r_s would take if harvesting were suddenly terminated. This rate is identical with the rate of non-selective harvesting that holds the population at $r_s = 0$.

For this situation a further rate of increase, r_p, is required. It is the 'potential rate of increase', defined as the exponential rate at which a population initially increases after one agent of mortality is eliminated.

The r_p of a harvested population can be measured directly by halting the harvesting, or calculated indirectly from a life table partitioned into rates of harvesting mortality and natural mortality (Section 8.3). It is more conveniently estimated from indices. The probability that an individual is included in a harvest has no close relationship either to its reproductive performance or to its ability to survive other agents of mortality. Consequently, indices such as fat reserves, fecundity and rate of natural mortality tend to correlate with r_p rather than with the rate r_s to which the population is artificially held by harvesting. When indices have been calibrated against r_s of an unharvested population they can be used as indices of r_p when a population is harvested. Partial calibration of several indices of this kind has been demonstrated for red deer (Caughley 1971b).

Chapter 6

Dispersal

Dispersal can be defined as the movement an animal makes from its point of origin to the place where it reproduces or would have reproduced if it had survived and found a mate (Howard 1960). It should not be confused with local movement (travelling within a home range) or with migration (movement between summer and winter home ranges). Dispersal can take bizarre forms— witness the occasional mass movements of the springbok *Antidorcas marsupialis* (Child and Le Riche 1969)—but more often it is a continuous unspectacular process detectable only by careful study.

The survival of a species is as dependent on dispersal as it is on reproduction and longevity. Habitats seldom remain stable for long and local populations therefore tend to be ephemeral, establishing, growing and extinguishing again as favourable habitat is created and destroyed. We can see this plainly when the habitat of a species comprises a single stage in a forest succession. The areas that can support local populations are scattered islands in a sea of unfavourable habitat. They are continually created by slips, fires and windthrow, and continuously destroyed by successional advance. As some populations are snuffed out by a change of habitat other populations spring up in areas entering the favourable stage of the succession. To survive, the species must contain at least some individuals capable of dispersing from one area of habitat to another.

Most local populations have a life cycle of this kind. They establish, grow and maintain themselves until an environmental change knocks them out again. The episode is telescoped when a population depends on a successional stage, and for that reason it is conspicuous, but the process is no less real for species with more generalized ecologies. It just takes longer. The habitat of a species is rarely continuous, being usually distributed as patches forming a set of elements in a mosaic of many habitats. Under the influence of climatic fluctuations, year to year variability in weather, plant succession, and the activities of man and other animals, the mosaic changes kaleidoscopically and with it changes the distribution of animal populations. Some patches of habitat increase in size, some shrink, and still others are created or destroyed. A population copes with these changes in three ways:

1. it can change genetically or phenotypically as the habitat changes, thereby allowing it to remain where it is;
2. it can be ecologically adaptable, able to cope with a broad range of habitat; or
3. some of its members must be able to disperse.

Each of these talents is a disability when conditions are relatively stable. The first demands that the population retains genetic variability, thereby condemning most of its members to carry a combination of genes sub-optimal in a stable environment. The second dooms population members to the role of earnest generalists. In being able to cope with a broad range of conditions they are seldom very efficient in any one set. The ability to disperse carries neither disadvantage, but dispersal is a wasteful process. It reduces a population's rate of increase, inhibiting a rapid build-up after density is lowered by a temporary misfortune. Gentry (1968) demonstrated this effect very well. He fenced areas containing pine mice (*Microtus pinetorum*), thereby holding within the population those mice that would otherwise have dispersed. The enclosed populations reached far higher densities than those whose dispersal was unrestricted.

Howard (1960) divided dispersal into two categories. 'Innate dispersal' is a spontaneous movement, the cause of which lies with the genetics of the dispersing individual rather than with environmental conditions. 'Environmental dispersal' is, in contrast, a behavioural response to unfavourable conditions such as a shortage of food or of space.

The two kinds of dispersal produce different results. In general, innate dispersal lacks a preferred direction; an animal at the edge of a population's range is as likely to disperse back into it as to colonize new territory. Environmental dispersal is less random. The dispersing individual tends to move away from unfavourable conditions, usually from a higher density to a lower density. Innate dispersal often entails a journey many times longer than the average radius of a home range whereas environmental dispersal more commonly entails a short journey ending immediately the animal encounters favourable conditions.

6.1 THE PROCESS OF DISPERSAL

A detailed study of dispersal requires this information:

1. the sex and age of dispersing individuals and the sex and age of individuals that do not disperse;
2. whether individuals disperse in random directions or whether the direction is influenced by density gradients, prevailing wind, angle of the sun, and so on;
3. whether dispersing individuals are hounded away from their place of birth or whether they leave of their own volition;

4. the probability that an individual will disperse, and whether this is a constant or a function of density, rate of increase, food supply, or some other influence;
5. the mean and variance of the distance moved, whether these are constants for a species or vary in response to environmental conditions, and whether the frequency distribution of dispersal distances is multimodal or unimodal;
6. the probability that a dispersing individual will survive to reproduce, as contrasted with this probability for an individual that does not disperse; and
7. the genetic differences between dispersing and sedentary individuals.

Generalizations on dispersal will not be possible until this information is gleaned from several species. So far it has not been collected from a single population (although the study of voles by Myers and Krebs (1971) comes close), a fact that is depressing but hardly surprising. Dispersal is the most difficult of all population processes to investigate, and for this reason it tends to be glossed over or ignored in most ecological studies. But the investigation of a population's ecology without reference to dispersal is analogous to a taxonomic study that ignores evolution. Both describe a dynamic process as if it were static. Studies of dispersal are laborious and technically difficult but the insight they provide is well worth the labour.

Although detailed studies of dispersal are lacking there have been several good studies on one or more aspects of the process.

Direction of dispersal

Frith's (1959, 1963) study on several species of ducks showed that individuals dispersed in all directions. However, he could not show from his band returns that directions were preferred equally because band were recovered by sportsmen whose density varied by direction.

Tanton (1965) found that wood mice, *Apodemus sylvaticus*, dispersed into his study area equally from all directions. He used the Rayleigh test (Durand and Greenwood 1958) to determine randomness of the angle between where each newcomer was first trapped and where it was trapped next. The test is simple:

$$\text{if } A_1, A_2, A_3, \ldots A_n \text{ are the angles,}$$
$$W = \sin A_1 + \sin A_2 + \sin A_3 + \ldots + \sin A_n, \text{ and}$$
$$V = \cos A_1 + \cos A_2 + \cos A_3 + \ldots + \cos A_n,$$

randomness of direction has a probability less than 5 per cent when $(W^2 + V^2)/n$ exceeds 3. If n is less than 6 the test breaks down and an exact probability method (Durand and Greenwood 1958) is used instead.

Caughley (1970b) showed that, in New Zealand, thar populations spread twice as rapidly north as south, an attribute they share with populations of chamois, *Rupicapra rupicapra*, in the same area (Christie and Andrews 1965, Christie 1967). The rate of dispersal might appear, therefore, to vary with direction. But equally likely these observations may reflect differences in rate of increase between populations established on north-facing as against south-facing slopes, interacting with a difference in the probability of successful dispersal across a valley as against across the ridge between valleys. As will be shown later (Section 6.3.2), the information provided by a population's pattern of spread is insufficient to reveal the behaviour of individuals contributing to the spread.

Sex and age of dispersing individuals

Howard and Childs (1959) captured dispersing pocket gophers, *Thomomys bottae*, in funnel traps located along drift fences. All of the 197 gophers caught in this way were sub-adults. The authors also caught a further 608 rodents divided between eight species. The majority of these were also sub-adults (Howard 1960). Caughley (1960) examined the remains of 28 crabeater seals, *Lobodon carcinophaga*, that had dispersed 25 km over rock and gravel in an ice-free valley in Antarctica. All were sub-adults. Several further examples of apparently innate dispersal of young birds and mammals led Howard (1960) to conclude that 'for the most part, the major dispersal movements are made by virgins about the time they attain puberty'.

Howard is probably right both in his conclusion and in his cautious qualification. Four studies have shown that it is unwise to assume, without investigation, that the spread of a given population is a simple consequence of juvenile dispersal. Stoddard (1950) found that most dispersing water voles, *Avicola terrestris*, were mature females, and Tast (1966) reached the same conclusion for bank voles, *Microtus oeconomus*. Caughley (1970b) calculated the age distribution of thar at the edge of breeding range and concluded that it was anomalous unless dispersing individuals included some mature females. Myers and Krebs (1971) found that dispersive behaviour was associated strongly with the attainment of sexual maturation, but that dispersal of adults was also common.

Genetics of dispersing individuals

Although the genetical entomologists can tell us something about the relationship between genetics and dispersal (e.g. Ford 1964, and the symposium edited by Baker and Stebbins 1965), little research has been aimed at sorting out this relationship for vertebrates. Hopefully, the paper by Myers and Krebs (1971) may signal the end of the drought.

They studied several characteristics of dispersing individuals of two species of voles, comparing these with the characteristics of sedentary individuals at four stages of an eruptive fluctuation.

Table 6.1 is a summary of their findings. Genotypes for two polymorphic

Table 6.1. Characteristics of dispersing voles, *Microtus pennsylvanicus*, at four stages of an eruptive fluctuation (Myers and Krebs 1971)

INCREASE
(1) Amount of dispersal high
(2) Dispersal accounts for high proportion of loss in control populations
(3) Female Tf–C/E phenotypes dispersing

EARLY PEAK
(1) Low proportion of young males dispersing
(2) High proportion of dispersing young males in breeding condition
(3) High proportion of young females dispersing
(4) Dispersing males show more threat and less avoidance behaviour
(5) Female Tf–C/E phenotypes dispersing
(6) Sex ratio of dispersing animals higher (more males)

LATE PEAK
(1) Male Tf–E and LAP–S phenotypes dispersing
(2) Dispersing males show more threat behaviour
(3) More light-weight females dispersing
(4) Exploratory behaviour high for all populations

DECLINE
(1) Amount of dispersal low
(2) Mortality not accounted for by dispersal
(3) Body weights of control and dispersing males low
(4) High proportion of young females dispersing
(5) High proportion of dispersing young females in breeding condition
(6) Low proportion of scrotal subadults in all populations
(7) Male Tf–E phenotypes dispersing

---

plasma proteins, leucine aminopeptidose (LAP) and transferrin (Tf), differed in frequency between samples of dispersing and sedentary individuals. One transferrin was found only in dispersing males. Further, the differences varied with the stage of the eruptive fluctuation. These results demonstrate that individual voles do not disperse for simple reasons, that the causes lie in a complex interaction between their genetics and their environment.

6.2 DETECTION OF DISPERSAL

Animal movements can be classified under four heads:

1. dispersal,
2. local movement within a home range,
3. nomadism,
4. migration.

Often we detect movement easily enough but strike difficulties in classifying it.

Migration can be differentiated from dispersal by marking juveniles where they are born. If these subsequently leave the area but return to breed, the movement is migratory. If they leave and are found breeding elsewhere the movement is dispersive.

Nomadism, itself a form of dispersal, is difficult to differentiate from local movement. By definition, nomadism implies the absence of a home range, and the two kinds of movement are therefore differentiated according to whether or not movement is restricted to a circumscribed area. A measurement of mean mobility, of itself, provides no answer; nomadic animals do not necessarily range over larger areas than those restricted to home ranges. Instead, nomadism must be detected from the relationship between range of movement and time. Nomadic movement is similar to a random walk (Section 6.3.1), a diagnostic feature of which is the tendency of an animal to be located, on average, farther and farther away from the place it was first seen as time goes on. If, however, an individual is restricted to a home range, the distance between where it was first seen and where it is seen subsequently tends first to rise by an increment proportional to the size of its home range and then holds constant with time.

In Figure 6.1 Frith's (1964) measurements of distance between points of marking and recovery of red kangaroos, *Megaleia rufa*, are plotted against time since marking. The regression slopes at 0·11 km/month which does not differ significantly from a slope of zero ($P = 0.6 - 0.7$). The observed movement of these marked animals suggests, therefore, that they were not nomadic but travelled within large but circumscribed home ranges.

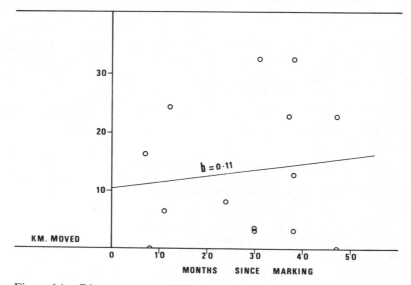

Figure 6.1.   Distances over which red kangaroos moved between marking and recapture (data from Frith 1964).

6.3 MODELS OF DISPERSAL

So far we have considered dispersal only in terms of individual animals. We will now consider how these movements, made independently by many individuals, affect the distribution of the population. In particular, we will investigate separately the net effects of innate dispersal and of environmental dispersal to determine how these two modes shape a population's distribution.

When dispersal is innate the proportion of the population that disperses is independent of the population's size and density. When dispersal is environmentally induced the proportion tends to rise, not necessarily smoothly, with increasing density. The difference can be summarized in the language of physics: innate dispersal is a process of diffusion; environmental dispersal is an effect of pressure. 'Pressure dispersal' does not imply that a dispersing individual is pushed away from its place of birth—it may well have left of its own accord—but that the result is the same as if it were.

6.3.1 Random walk

Suppose a jerboa fell into a narrow ditch and bounded in panic backwards and forwards along it. If it jumped once every second, and if each jump carried it one metre, and if the direction of each jump were random, we could calculate the probability of the jerboa being at a given position at any given subsequent time. The point at which it fell is arbitrarily designated 0 and metres are measured from there. One m to the left is therefore -1 and one to the right is $+1$.

Table 6.2 shows the distribution of these probabilities around the origin.

Table 6.2. Probability of the location of a jerboa that jumps one metre each second in random directions

Time in seconds	Metres left and										right of origin			Variance in metres
	-6	-5	-4	-3	-2	-1	0	1	2	3	4	5	6	
0							1							0
1						$\frac{1}{2}$		$\frac{1}{2}$						1
2					$\frac{1}{4}$		$\frac{2}{4}$		$\frac{1}{4}$					2
3				$\frac{1}{8}$		$\frac{3}{8}$		$\frac{3}{8}$		$\frac{1}{8}$				3
4			$\frac{1}{16}$		$\frac{4}{16}$		$\frac{6}{16}$		$\frac{4}{16}$		$\frac{1}{16}$			4
5		$\frac{1}{32}$		$\frac{5}{32}$		$\frac{10}{32}$		$\frac{10}{32}$		$\frac{5}{32}$		$\frac{1}{32}$		5
6	$\frac{1}{64}$		$\frac{6}{64}$		$\frac{15}{64}$		$\frac{20}{64}$		$\frac{15}{64}$		$\frac{6}{64}$		$\frac{1}{64}$	6

The numerators form Pascal's triangle, each numerator being the sum of the two flanking it in the row above. The probabilities are symmetrical around the origin and the distribution broadens progressively, variance being related directly to time. Had 100 jerboas been dumped into the ditch, the same pattern would result if they did not interfere with each other's movements. After five seconds about $31 = 100 \times 10/32$ would be one metre right of the origin, 15 three metres right and three at five metres right. The same frequencies would be expected left of the origin.

This improbable model illustrates the characteristics of diffusion along one dimension:

1. the most likely position at any one time is the origin,
2. the distribution of probabilities is symmetrical around the origin,
3. the variance of the distribution expands steadily with time, and
4. the rate of spread is constant with time.

These attributes are the essence of a random walk in one, two or three dimensions. If we take any diameter through the point of liberation, irrespective of the number of dimensions within which dispersal occurs, the positional probabilities along this line will conform to the above rules.

6.3.2 Complex models

The pseudo-jerboa model is the simplest case of dispersal by diffusion. It must be elaborated considerably before it begins to mimic the dispersal of real animals. If a population were initiated by liberating several individuals at the centre of a homogeneous area, the dispersive behaviour of these animals would differ from that of pseudo-jerboas in these respects:

1. only a small number would disperse,
2. the distance dispersed would vary between individuals,
3. since the animals reproduce, the population's size, and hence the number of animals that disperse, increases with time, and
4. animals disperse in at least two dimensions.

The mathematics of this more complex process were investigated by Fisher (1937), Kendall (1948), Skellam (1951, 1955) and Pielou (1969: 129) who concluded that the complexity has little effect on the outcome: the radius of distribution increases linearly with time.

Figure 6.2 graphs the spread in one dimension of a population simulated in a computer. The population is initiated with ten animals that reproduce at a yearly rate of $r = 0.1$. Each year one in every ten animals, on average, disperses a mean distance of one km with a standard deviation of 0.4 km. The direction of dispersal is random, the dispersing animals and the directions in which they move being chosen by lottery. The inset of Figure 6.2 graphs the

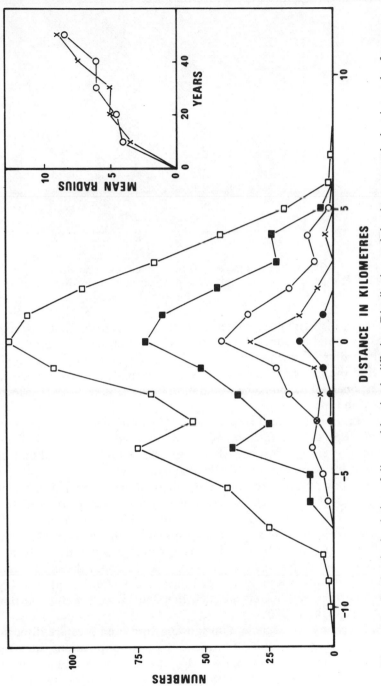

Figure 6.2. Stochastic simulation of dispersal by simple diffusion. Distribution and density are graphed at intervals of ten years.

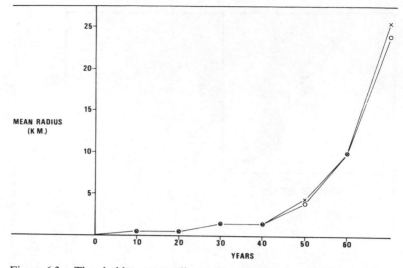

Figure 6.3. Threshold-pressure dispersal; a stochastic simulation of the spread of a population (two replicates) in which individuals disperse only when local density passes a critical threshold.

average radius of distribution at intervals of ten years, and also that of a replicate population expanding under the same rules. The regression is approximately linear, the radius lengthening by about the same increment in each ten-year period. Had the pattern of population growth been defined as logistic rather than as exponential the outcome would have been the same—a linear increase of radius with time.

The antithesis of dispersal by diffusion is pressure dispersal. Suppose that, instead of dispersing in random directions at a constant rate, individuals disperse only when the density reaches some critical threshold, and they move only so far as the nearest area below threshold density.

Figure 6.3 shows the outcome of a stochasticized simulation of this threshold-pressure process. Two replicate populations are each launched with ten animals that subsequently increase at $r = 0.1$ per year. The pattern of spread, as summarized by the trend of mean radius on time, contrasts with that resulting from diffusion (Figure 6.2). Populations spreading under the influence of threshold-pressure dispersal hang fire until the threshold density is reached in the area of liberation, and then extend their radius of distribution each year by a steadily increasing increment such that the trend of radius on time is exponential (Caughley 1970b).

These two examples—simple diffusion and threshold-pressure dispersal—are the two extremes of a range of processes whereby populations extend their distributions. Neither extreme is likely to be common in nature. Dispersal by simple diffusion may be an adequate description of the movement of gas molecules but the movement of social animals is unlikely to be so mechanical. The strictly directional dispersal generated by a threshold-pressure model is also an exaggeration of what happens in nature. Individuals would not be expected to move unerringly towards the closest area below threshold density, although their movements may well be influenced by local density gradients. Likewise, since the two models each describe the effects of only one kind of dispersal—innate dispersal in the diffusion model and environmental dispersal in the threshold-pressure model—they exclude any process comprising a mixture of the two modes.

A model striking a compromise between the two extremes is much more likely to mimic natural dispersal. Suppose that the proportion of individuals dispersing each year is independent of density (as in the diffusion model) but that a dispersing individual moves away from the area of highest density in its immediate vicinity (as in the pressure model). Figure 6.4 graphs the outcome of such a process, giving the regression of radius on time for the distribution of two model populations, each with two replicates. Both populations increase at the rate $r = 0.1$ per year and on average 10 per cent of their members disperse each year. Dispersing individuals of the first population move an average of 2 km, and in the second, 1 km. Despite the directional dispersal built into this model the mean radius of distribution is approximately linear on time, as it is for the diffusion model. The process modelled here cannot be differentiated by outcome from the process of diffusion.

More generally, it is easily shown by computer simulations that whether the proportion of animals dispersing each year is either a constant or an increasing function of density, whether the direction of dispersal is either random or away

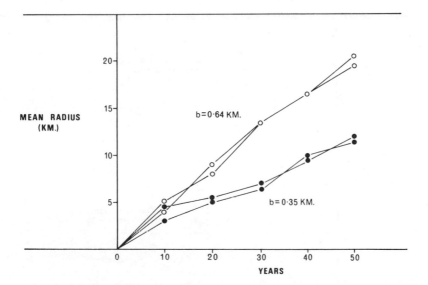

Figure 6.4. Stochastic simulation of the spread of two populations each with two replicates. Dispersing individuals of the first move an average of 2 km, and of the second, 1 km. The mode of dispersal includes elements of both diffusion and pressure.

68

from local high density, and whether population growth is either logistic or exponential, the population spreads at an approximately constant rate and the process cannot therefore be differentiated by outcome from that of the diffusion model. The simulations described here emphasize the difficulties involved in deducing the dispersive behaviour of individuals from the effect of their behaviour on a population's range of distribution.

6.4 PATTERNS OF SPREAD

Figure 6.5 diagrams the expansion of two populations initiated by liberation. The first graph tracks the spread of muskrats, *Ondatra zibethica*, introduced into Central Europe in 1905, the distribution being mapped on five occasions over the next 25 years. The second summarizes the spread through the Southern Alps of Himalayan thar introduced into New Zealand in 1904, the distribution being mapped in 1936, 1946, 1956, and 1966. In both cases the area of occupation is expressed as a radial equivalent, the radius of a circle with the same area as the mapped distribution.

Muskrats spread at an approximately constant rate, a pattern that has been interpreted (Skellam 1951, 1955; Caughley 1970b) as a product of simple diffusion. However, the results of computer simulations reported in Section 6.32 indicate that such a facile conclusion is not justified. The spread of muskrats is certainly inconsistent with threshold-pressure dispersal but the regression cannot be used to differentiate between simple diffusion and directional

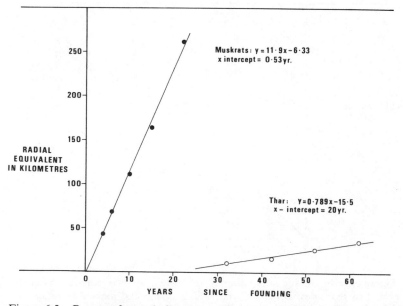

Figure 6.5. Pattern of spread of muskrats liberated in Europe and Himalayan thar liberated in New Zealand (data from Skellam 1955 and Caughley 1970b).

dispersal, between a process involving a constant proportion of dispersers or a proportion rising with density, or between combinations of these processes, since each generates an approximately linear trend of radial equivalent on time.

The spread of the thar population suggests an even more complex history. Thar spread at an approximately constant rate for the 30 years over which distributions were mapped, but an extrapolation of this trend backwards demonstrates that previously the rate of spread was lower. This pattern would suggest threshold-pressure dispersal were it not for the constancy of the increment added to the radial equivalent each decade between 1936 and 1966. Although the trend could be 'explained' in a number of ways, in the absence of information on the behaviour of individual animals one explanation is as valuable or as worthless as another.

The trend of radial equivalent on time does not reveal much about the processes by which populations spread. It may allow rejection of extreme hypotheses but it does not differentiate between a number of very different processes.

6.4.1 Rate of spread

The speed at which a population extends its range can be summarized in a number of ways. Only two are used here. If distribution is mapped on two occasions the spread in any direction can be measured as the shortest distance between the second boundary and a given point on the first. The greatest of all such distances is the maximum spread between the two occasions and this distance is divided by the time lapse to give the maximum rate of spread for that period. Alternatively, a mean rate of spread is calculated as the increase in the radial equivalent ($\sqrt{\text{area}/\pi}$) of range between the two occasions, divided by the time lapse.

Both methods have disadvantages. The first provides a rate that may be highly atypical of the general rate of spread in the period under review. The second makes biological sense only when movement is unrestricted in all directions. If a population's range abuts the sea or a high mountain the population cannot spread farther in that direction; mean rate of spread is then an abstraction underestimating rate of spread in other directions. Since the spread of most populations is impeded somewhere along the boundary, the first method usually allows a more precise comparison between species. Table 6.3 compares the spread of several species in terms of maximum rates.

These estimates are rough. They should not be interpreted as constants for the listed species but as rates of spread over specific periods in specific areas; rate of spread is strongly dependent on environmental conditions. But in a general sense these estimates allow some comparison between species. They demonstrate that species differ markedly in their ability to extend breeding range and that this ability is correlated only weakly with average mobility within home ranges. Few would predict, for instance, that muskrats can spread several times faster than white-tailed deer.

Table 6.3. Rates of spread

Species	km/yr	Data from:
European rabbit	64	Myers 1970
Horse	48	Darwin 1845
Masked shrew	19	Buckner 1966
Muskrat	15	Skellam 1955
Cotton rat	12	Cockrum 1948
Chaffinch	11	Kalela 1949
Primitive man	11	This section
Nine-banded armadillo	10	Humphrey 1974
European blackbird	9·2	Kalela 1949
Chamois	8·7	Caughley 1963
European polecat	5·1	Kalela 1949
Grey kangaroo	4·8	Caughley unpublished
Himalayan thar	3·2	Caughley 1970b
Red deer	1·6	Caughley 1963
Sika deer	1·6	Caughley 1963
Whitetailed deer	1·0	Caughley 1963
Fallow deer	0·8	Caughley 1963
Rusa deer	0·8	Caughley 1963
Sambar deer	0·6	Caughley 1963
Wapiti	0·6	Caughley 1963

The rate of spread of primitive man can only be guessed at but an approximate estimate is possible from what is known of the colonization of the Americas. Several estimates of the date at which men entered Alaska appear in the archaeological literature. They are centred on about 10,000 B.C., although a few radio-carbon dates (of hotly disputed validity) argue for a much earlier arrival. The earliest date that currently can be placed on man's arrival in southern South America comes from artefacts returning a radio-carbon date of about 8,500 B.C. (Bird 1970, Martin 1973). If these two dates can be trusted they imply a mean rate of spread of 11 km/yr. Even if this rate is in error by as much as double or half, as it well might be, it indicates that the rate of spread of primitive man is within the range of rates characterizing other mammals. Man may be unique, but not with respect to his rate of spread.

Feral horses spread much more rapidly. A group liberated at Buenos Aires in 1537 founded a population that had extended to the Straits of Magellan by 1580 (Darwin 1845 : 222): a rate of spread of 48 km/yr.

Whether or not it teaches us anything about the spread of populations, it is difficult to avoid speculating on how fast a species would spread if it combined the brains of man with the stamina of the horse. How fast do men spread when they are mounted and how fast do horses spread when they are ridden? A rough estimate is possible from the rate of expansion of the Mongol empire, first under Genghis Khan and later under his son Ogodai. The western campaign was launched in 1219 A.D. and continued until 1242 when Batu and

Subetai, the two brilliant field generals of the European advance, called off the thrust into western Europe on receiving news of Ogodai's death (Lister 1969). At this time units of the Mongol cavalry were probing the outer defences of Vienna, 7,000 km from their homeland south of Lake Baikal. The rate of spread was an astonishing 300 km/yr. Roughly then, horsemen spread six times faster than riderless horses and thirty times faster than men on foot.

Chapter 7

Fecundity

The fecundity rate of a female is measured as the number of live births she produces over an interval of time, generally one year. In population biology the fecundity rate of a given female is trivial information; the data of interest are the mean fecundity rates of each female age-class. When fecundity is used in analysis it is expressed as the mean number of *female* live births per female over an interval of age. A complete schedule of these values, covering all ages from birth to the oldest observed age, is called a fecundity table. Before it can be constructed we need to know for each age class of mothers the mean litter size, the mean number of litters produced per year and the sex ratio at birth.

The way a fecundity table is used in analysis depends on whether the population's breeding system is best approximated by the birth-pulse or the birth-flow model. The initial aim of a study on fecundity is therefore to establish the mean or median date and the standard deviation of the season of births.

7.1 SEASON OF BIRTHS

The seasons impose an annual cycle of breeding activity on most populations. Although the cycle has no natural beginning or end, an arbitrary beginning must be specified to allow division of the population into age classes. The most convenient 'beginning' is the average date of birth. Without much loss of accuracy the ages of all individuals in a birth-pulse population can be expressed on this day as a whole number of years.

Most published descriptions of a season of births are imprecise. They usually give an observed spread of birth dates and the 'peak period' of births. In my experience, subjective estimates of 'peak period' overestimate the modal date of birth, and estimates of spread tell little about the season's dispersion. Vague parameters must be discarded if the season of births is to be described precisely. The average birthday should be calculated as the mean or median of one or more seasons, and the dispersion of the season should be measured by the standard deviation of the dates of birth.

Direct method

A straightforward method of calculating these statistics is to divide the season of births into equal periods and to count the births occurring in each. This method is used only when a population, or a sample of it, is under continual observation.

Table 7.1 gives a set of data amenable to direct analysis. It is taken from Brand's (1963) records of 103 Himalayan thar born in the zoo at Pretoria between 1908 and 1959. These data were reported as a histogram of births per half month. Two additional births recorded by Brand, one in late March and one in late April, are excluded from Table 7.1.

The mean date of birth and the standard deviation of the season are first calculated in terms of the period codes:

$$\text{Mean date} = \frac{\sum fx}{\sum f} = \frac{338}{103} = 3 \cdot 28$$

$$\text{Variance} = \frac{\sum fx^2 - (\sum fx)^2 / \sum f}{(\sum f) - 1}$$

$$= \frac{1261 - 338^2 / 103}{102} = 1 \cdot 49.$$

Because this variance has been calculated from data grouped into intervals of constant width it is a slight overestimate that is corrected by subtracting a twelfth of the squared interval (Sheppard's correction for grouping) to give

$$\text{Variance} = 1 \cdot 49 - \frac{1}{12} = 1 \cdot 41$$

since the interval has a width of one unit in its coded form. The standard deviation is estimated as the square root of this variance: $s = \sqrt{(1 \cdot 41)} = 1 \cdot 19$. The

Table 7.1.   Frequency of births per half month for thar in the Pretoria Zoo (Brand 1963)

Half month	Period code x	Number of births f	fx	fx^2
Oct. a	0	0	0	0
Oct. b	1	5	5	5
Nov. a	2	22	44	88
Nov. b	3	39	117	351
Dec. a	4	17	68	271
Dec. b	5	17	85	425
Jan. a	6	2	12	72
Jan. b	7	1	7	49
Feb. a	8	0	0	0
		$\sum f = 103$	$\sum fx = 338$	$\sum fx^2 = 1261$

results are now transformed from period codes to real dates:

$$\text{Mean} = 3\cdot28 \text{ periods } (1 \text{ period} = 15\cdot2 \text{ days})$$
$$= 50 \text{ days after mid-point of period } 0$$
$$= 7 \text{ October} + 50 \text{ days}$$
$$= 26 \text{ November.}$$

$$\text{Standard deviation} = 1\cdot19 \text{ periods} \times 15\cdot2$$
$$= 18\cdot1 \text{ days}$$

$$\text{Standard error} = 18\cdot1/\sqrt{(103)}$$
$$= 1\cdot8 \text{ days.}$$

Most seasons of birth are skewed in the positive direction, i.e. the frequency of births climbs rapidly to a peak and then tails off more slowly. In this case the median rather than the mean date of birth is a better indication of the average birthday of all animals in the population. Mortality in the first year of life acts more heavily on the late births, forcing the average birthday of the survivors backward from the mean towards the median. Median date of birth is calculated (Simpson *et al.* 1960) as:

$$M = L + \frac{i - \frac{1}{2}}{f}$$

where L = the true lower limit of the class in which the sample median lies,
$\quad\ i$ = the serial number of the desired observation within the class, and
$\quad\ f$ = the absolute frequency of the median class.

For the example in Table 7.1,

$$M = 2\cdot5 + \frac{23 - \frac{1}{2}}{39}$$

$$= 3\cdot08 \text{ periods}$$
$$= 23 \text{ November.}$$

Indirect method A

The direct method is applicable only to animals in captivity, and to intensively studied breeding colonies. Its use is limited by the practical difficulties of estimating the proportion of births produced during successive intervals of time. Easier to obtain is an estimate of the number of births produced prior to a specific date, expressed as a proportion of total births produced over the entire season. Indirect estimates of the statistics calculated previously can be made from these data. Figure 7.1 shows the relationship between 'births per period', the data analysed by the direct method, and 'births to date' which provide data for indirect analysis.

Table 7.2 provides data of this kind. Again, the species is the Himalayan thar, but this time the data are from a free-ranging population in the mountains

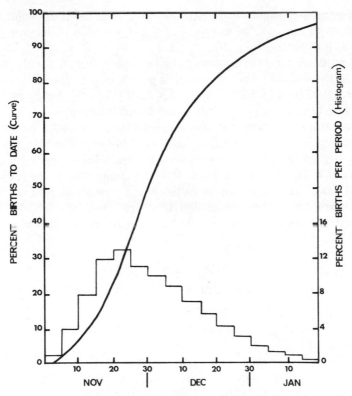

Figure 7.1. Season of births of Himalayan thar graphed as a cumulative curve giving the percentage of births up to a given date, and as a histogram of per cent births per interval (Caughley 1971a).

Table 7.2. The reproductive status of female Himalayan thar at six dates during the season of births (Caughley 1971a)

Date	Day	Log day	Number pregnant	Number lactating	Total	% births to date	Probit %
	X	x	P	L	B	$100L/B$	y
Nov. 17	1	0·00	23	3	26	12	3·83
Nov. 22	6	0·78	18	6	24	25	4·33
Dec. 12	26	1·41	12	26	38	68	5·47
Jan. 6	51	1·71	2	15	17	88	6·18
Jan. 31	76	1·88	3	49	51	94	6·56
Feb. 5	81	1·91	0	34	34	100	—

of New Zealand. The table lists for five dates during the season of births the numbers of pregnant and of lactating females shot during five-day periods centred on these dates. The lactating females had given birth before they were shot; the pregnant females had yet to do so. Hence the proportion of births produced up to the date of sampling is the proportion of lactating females sampled at that date.

Sampling dates are converted to days beginning from 1 for the first date. Each "% births to date' is transformed to a probit, tables of which are provided by Finney (1947) and by Fisher and Yates (1948). The 5th February sample is discarded because the season of births had ended by then.

A probit transformation converts a sigmoidal curve of "% births to date' into a straight line if the underlying season of births is normally distributed. The first step of the analysis is therefore to check for linearity by graphing probits against day of observation, X. If the resultant trend is convex upwards, as in this case, the distribution of births has a positive skew. It can usually be normalized for further calculation by transformation X to log X, here symbolized x. A regression is now calculated for probit (y) against x:

$$n = 5$$

$$\bar{x} = 1\cdot156 \qquad \bar{y} = 5\cdot274$$

$$\sum x = 5\cdot780 \qquad \sum y = 26\cdot370$$

$$\sum x^2 = 9\cdot055 \qquad \sum xy = 33\cdot991$$

$$(\sum x)^2/n = 6\cdot682 \qquad (\sum x)(\sum y)/n = 30\cdot484$$

$$SS_x = \sum x^2 - (\sum x)^2/n = 2\cdot373 \qquad SS_{xy} = \sum xy - (\sum x)(\sum y)/n = 3\cdot507$$

From SS_x, the sum of squares of x, and SS_{xy}, the sum of products of x and y, the slope b of the regression line and the y-intercept a are calculated;

$$b = SS_{xy}/SS_x = 1\cdot48$$
$$a = \bar{y} - b\bar{x} = 3\cdot56$$

The median date of birth is calculated by

$$'M' = (5 - a)/b = 0\cdot97,$$

but this is the median on the day scale transformed to logs. The true median is its antilog, 9, which is 26th November.

The standard deviation of the season of births is the reciprocal of the regression slope:

$$'S' = 1/b = 0\cdot68 \text{ log days.}$$

An approximation to the true standard deviation is

$$s = \frac{\text{antilog}('M' - 'S') - \text{antilog}('M' + 'S')}{2}$$

$$= \frac{\text{antilog}(0\cdot97 - 0\cdot68) - \text{antilog}(0\cdot97 + 0\cdot68)}{2}$$

$$= 21\cdot5 \text{ days}$$

The regression method outlined here is a simplification of a more complex method termed probit analysis. Although simple regression is accurate enough for most purposes, probit analysis should be used when samples differ markedly in size or when maximal accuracy is sought. Finney (1947) gives a lucid explanation of the method. Data in Table 7.2 are a selection, for purposes of illustrating the method, of a larger body of data (Caughley 1971a). Probit analysis of the total data returned a median of 30th November and a standard deviation of 18·5 days.

The direct method of analysis required frequency of births in equal intervals of time. In contrast, this indirect method requires information on births up to a specific date. Ideally, the samples recorded in Table 7.2 should each have been shot over one day. When, as in this case, such intensive sampling is not practicable, the periods should be kept as short as possible.

This method allows great flexibility of experimental design. It has these advantages over the direct method:

1. Sampling periods need not be contiguous.
2. The interval between sampling periods may vary.
3. Numbers per sample, and hence effort expended, need not be the same.
4. The calculated '% births to date' are statistically independent of each other. Loss of data from one period does not bias the calculated median and standard deviation.
5. Neither the beginning nor the end of the season of births need be sampled, a distinct advantage when these dates are unknown.

A modification of this method, which circumvents the need to capture or kill animals, is given by Caughley and Caughley (1974).

Indirect method B

Ornithologists usually date the reproductive season by counting nests containing eggs at several dates throughout the season, equal effort being expended on searching at each date. When graphed against date of sampling these totals usually show a positive skew (Davis 1955). The distribution of numbers of nests discovered at different dates is not the same as the distribution of dates of laying or of hatching, although it has often been treated as such. The 'nesting distribution' has a mean date beyond the mean date of laying, a lesser skew than the distributions of laying dates and of hatching dates, and a standard deviation greater than either.

A nesting distribution can be converted to a distribution of laying dates by stripping off that part of its variance originating from the length of the incubation period. When counts are made at regular intervals, the total for each date can be treated as a frequency centred at that date. The frequencies make up a distribution whose class interval is the time between searches. A variance s_n^2, expressed in units of the class interval, and a mean date $\bar{x}_n$, is calculated from the distribution, the subscripts indicating that these statistics are of the nesting distribution, not of the distribution of laying dates.

Next, the variance due to the incubation period is calculated. Suppose incubation lasts 20 days and the class interval is five days. The incubation period therefore has a length z of four class intervals. The variance due to incubation is the variance of the numbers 1, 2, 3 and 4, i.e. $s_z^2 = 1\cdot667$.

An estimate of the standard deviation of laying dates s is given by

$$s = \sqrt{s_n^2 - s_z^2}$$

in units of the class interval. The mean date of laying $\bar{x}$ is estimated from the mean of the nesting distribution and the mean length of the incubation period as

$$\bar{x} = \bar{x}_n - \frac{z+1}{2}$$

A mean date of hatching can be calculated as the mean date of laying plus the mean incubation period.

A reasonably accurate but slightly underestimated standard deviation of hatching dates is the calculated standard deviation of laying. The unbiased estimate is the square root of an addition of the variance of laying and the variance of the incubation period. The latter variance is not s_z^2 calculated above but a variance of several observed periods of incubation.

This method need not be restricted to birds and reptiles. It works equally well on mammals. The appropriate mammalian data are the proportions of adult females pregnant in samples taken regularly throughout the season of gestation. Sampling effort need not be constant by period. The method is particularly appropriate to small mammals that breed more than once during the season, and which cannot therefore be treated by indirect method A.

7.2 FREQUENCY OF BIRTHS

The mean number of litters produced per female over a year is estimated in two steps. The first requires an estimate of the mean prevalence of pregnancy over that portion of the year that pregnancies can be found. Prevalence of pregnancy, the proportion of females pregnant at a particular time, is measured on several occasions during this period. These estimates are averaged to give $\bar{P}$, the mean prevalence. For example, suppose that pregnant females can be found over five months of the year and that a sample is taken in each of these months. The five ratios of pregnant to total females are, in order,

$$\frac{3}{30}, \quad \frac{24}{48}, \quad \frac{12}{30}, \quad \frac{12}{40}, \quad \frac{7}{70}.$$

Mean prevalence of pregnancy over this period is therefore

$$\bar{P} = \frac{0\cdot1 + 0\cdot5 + 0\cdot4 + 0\cdot3 + 0\cdot1}{5} = 0\cdot28$$

P is an unweighted mean; it is not calculated by dividing the sum of the numerators of the monthly ratios by the sum of their denominators (the weighted mean).

Davis and Golley (1963 : 214) showed that the mean incidence of pregnancy, I, the number of times an average female becomes pregnant during a year, can be estimated from $\bar{P}$. First we define $\bar{D}$, the mean duration of visible pregnancy, as the mean length of gestation in which pregnancy can be detected. It will always be shorter than the true length of gestation because of the difficulty of confirming pregnancy in its initial stages. $\bar{D}$ is measured as a fraction of the length of the season over which pregnancies were sought. Continuing the example, we will assume that pregnancy can be detected over the last month of gestation. Since pregnant females were sought over five months of the year, $\bar{D} = 1/5 = 0.2$. The incidence of pregnancy can now be calculated as

$$\bar{I} = \bar{P}/\bar{D}$$
$$= 0.28/0.20$$
$$= 1.4,$$

indicating that an average female produces 1·4 pregnancies per year.

The accuracy of $\bar{I}$ is not affected by overestimating the length of the season of gestations. Suppose we had taken two further samples to estimate prevalence, one before the rutting season and one after the season of births. Each would return a zero prevalence of pregnancy. The mean prevalence of pregnancy would then be estimated over the seven months as

$$\bar{P} = \frac{0 + 0.1 + 0.5 + 0.4 + 0.3 + 0.1 + 0}{7} = 0.2$$

and the mean duration of visible pregnancy as

$$\bar{D} = 1/7 = 0.143.$$
$$\text{Hence } \bar{I} = 0.2/0.143 = 1.4,$$

the same result as obtained previously.

Strictly speaking, $\bar{I}$ is not the mean number of litters produced per year per female but the number of litters carried to the midpoint of visible pregnancy. If some litters are lost by abortion or resorbtion $\bar{I}$ will overestimate the average number of live litters produced.

We usually estimate $\bar{I}$ to allow calculation of a fecundity table that lists production of offspring for each age class of mother. $\bar{I}$, therefore, must be estimated for each age. D can safely be assumed independent of age but $\bar{P}$ may vary considerably between age classes. Consequently, the samples collected to determine prevalence of pregnancy must be stratified by age, $\bar{P}$ being estimated for each age class separately to allow calculation of age-specific values of $\bar{I}$.

7.3 SEX RATIO

Most vertebrates have a sex ratio at birth so close to 1 : 1 that the usual

Table 7.3. Test for disparate sex ratio of kangaroo pouch young (Caughley and Kean 1964)

Species	Sex	Observed number O	Expected number E	$O - E$	$\dfrac{(O - E)^2}{E}$	P
Red	Males	202	410/2	3	0·004	
Kangaroo	Females	208	410/2	3	0·004	
		410			$\chi^2 = 0·008$	0·95 (d.f. = 1)
Grey	Males	242	420/2	32	4·876	
Kangaroo[a]	Females	178	420/2	32	4·876	
		420			$\chi^2 = 9·752$	0·005 (d.f. = 1)

[a] This sample is from south-west Queensland. The sex ratio of Poole's (1973) sample of 555 pouch young from New South Wales does not differ significantly from that presented here, but Kirkpatrick's sample of 260 from south-central Queensland has a significantly different sex ratio.

slight preponderance of males can be ignored. If a marked disparity of sexes is suspected the observed sex ratio should be tested by χ^2.

Table 7.3 demonstrates the test on pouch young of grey kangaroos and red kangaroos. The ratio of sexes in the red kangaroo sample is not significantly different from the expected 1 : 1; the χ^2 probability for the grey kangaroo sample is low enough to suggest that the observed disparity in favour of males is real, not just a sampling error.

When sex ratios are used in population analysis they are expressed as the proportion of females, P_f. Hence P_f for red kangaroo pouch young is 0·50, and that for grey kangaroos is 0·42. The standard error of P_f is

$$SE = \sqrt{P_f(1 - P_f)/n},$$

where n is the number of males and females in the sample. For grey kangaroos

$$S.E. = \sqrt{(0·42 \times 0·58/420)}$$
$$= 0·024,$$

and the 95 per cent confidence limits equal $P_f \pm 2$ S.E. $= P_f \pm 0·048$. This value indicates that in 95 out of 100 samples, each of 420 pouch young, the observed proportion of females would fall between 0·372 and 0·468.

Sex ratios are sometimes given as 'males per hundred females'. By this convention the sex ratio of grey kangaroo pouch young is

$$\frac{242 \times 100}{178} = 136$$

with 95 per cent confidence limits (Riney 1956) of

$$\pm 200 \sqrt{\left(\frac{Mn}{F^3}\right)}$$

$$= \pm 200 \sqrt{\left(\frac{242 \times 420}{178^3}\right)}$$

$$= \pm 27.$$

7.4 FECUNDITY TABLE

The fecundity of an animal is measured as the number of offspring produced over an interval of age. Mammalian fecundity is measured as production of live births. For fish, reptiles and birds it is measured as the number of eggs produced, and less commonly as the number hatching. Because fecundity changes with age, a complete description of reproductive performance requires a separate calculation for each interval of the life span. The age interval is set at one year for birth-pulse populations and usually at less than one year for animals that produce more than one litter annually.

The fecundity pattern of a population is tabulated as a 'fecundity table' or 'fecundity schedule' listing the mean fecundity of animals in each age class. Two tables could be produced: one for each sex. Only one of these is required for population analysis, and because the fecundity of females is easier to measure the female fecundity table is always chosen. As a further simplification, only female offspring are recorded. The fecundity table used in population analysis lists the mean number of female offspring produced per female for each interval of age.

This statistic is designated m_x for a birth-pulse population, the subscript

Table 7.4. Fecundity tables for two populations of red deer (Lowe 1969 and Caughley 1971b)

Age in years x	Female live births per female	
	Rhum m_x	New Zealand m_x
0	0·000	0·000
1	0·000	0·000
2	0·000	0·063
3	0·311	0·415
4	0·278	0·400
5	0·302	0·455
6	0·400	0·414
7	0·476	0·486
8	0·358	0·476
9	0·447	0·455
10	0·289	0·500
> 10	0·283	0·375

referring to age. Thus m_3 is the mean number of female offspring produced at the birth pulse by three-year-old females. The analogous statistic for birth-flow populations is written $m_{x,x+1}$, being the mean number of female offspring produced per female over the *interval* starting at age x and ending one unit of age later. When one year is the unit of age employed, $m_{2,3}$ is the fecundity rate between a female's second and third birthday.

Table 7.4 is a fecundity table for two populations of red deer *Cervus elaphus*, one living on the island of Rhum off the coast of Scotland, the other living in the extreme south of New Zealand. When fecundity is listed in this way the fecundity pattern of the population can be taken in at a glance. No elaborate analysis is required to show that the New Zealand population has a generally higher fecundity than the deer living on Rhum.

Slightly different methods are used to estimate the fecundity schedules of birth-pulse and birth-flow populations. An example will be given of each.

Birth-pulse fecundity

Table 7.5 gives the data needed to calculate a fecundity schedule of chamois *Rupicapra rupicapra*. A sample of 275 females one year of age or more was shot during the season of births. Each animal was scored as breeding during the season or barren for that season, according to whether it was pregnant or lactating, or showing no sign of breeding activity. The sex ratio at birth is taken as 1 : 1, and litter size as one.

Table 7.5. A fecundity schedule calculated for chamois (Caughley 1970c)

Age in years x	Sampled number f_x	Number pregnant or lactating B_x	Female births/female $(B_x/2f_x)$ m_x
0	—	—	0·000
1	60	2	0·033
2	36	14	0·194
3	70	52	0·371
4	48	45	0·469
5	26	19	0·365
6	19	16	0·421
7	6	5	0·417
> 7	10	7	0·350

Birth-flow fecundity

Table 7.6 demonstrates calculation of a fecundity schedule for Orkney voles *Microtus orcadensis* kept in a laboratory. In these conditions the species breeds throughout the year. Leslie *et al.* (1955) reported the breeding histories of 51 females. Since some females were not continuously housed with males, the average number exposed to mating during a given interval is fractional.

Table 7.6. A fecundity schedule calculated for Orkney voles (Leslie *et al.* 1955)

Age in intervals of six weeks[a] $x, x+1$	Mean number of females $f_{x,x+1}$	Number of live young $B_{x,x+1}$	Female births/female $(B_{x,x+1}/2f_{x,x+1})$ $m_{x,x+1}$
0, 1	38·5	30	0·390
1, 2	43·0	114	1·326
2, 3	45·0	135	1·500
3, 4	44·5	137	1·539
4, 5	43·5	142	1·632
5, 6	43·0	142	1·651
6, 7	42·0	145	1·726
7, 8	40·0	139	1·737
8, 9	40·0	94	1·175
9, 10	40·5	86	1·062
10, 11	39·0	72	0·923
11, 12	38·0	42	0·553
12, 13	36·5	29	0·397
13, 14	31·0	22	0·355
14, 15	20·0	13	0·325
15, 16	12·5	6	0·240
16, 17	12·5	6	0·240

[a] Age 0 = 9 weeks after birth

7.5 SIMPLIFYING A FECUNDITY TABLE

The fecundity rate of most mammals and birds climbs from puberty and then levels off. The 'plateau' is invariably convex with the degree of curvature varying between groups. Small mammals and birds show a decline in fecundity after middle age, but deviation from a constant rate of adult fecundity is so slight for large mammals (Figure 7.2) that large samples are needed to detect

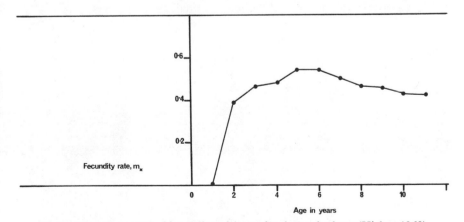

Figure 7.2.   The trend of fecundity with age for domestic sheep (Hickey 1960).

the curvature. For the purposes of population analysis, fecundity rate at the plateau can be expressed as a mean value unless curvature is extreme. However, if the aim is to determine the exact relationship between fecundity and age, as it might be in a study of reproductive physiology, a full fecundity table is required.

The average of the fecundity rates over the ages where fecundity is approximately constant can be calculated as either a weighted or unweighted mean. A weighted mean for ages equal to or greater than three years is calculated as

$$\bar{m}_w = \frac{n_3 + n_4 + n_5 \ldots + \text{etc.}}{N_3 + N_4 + N_5 \ldots + \text{etc.}},$$

n being the number of female offspring produced by N females of specified age. So long as no female is younger than the lower limit of the range of ages covered by the mean, the sampled females need not be aged.

An unweighted mean takes the form

$$\bar{m}_u = \frac{\dfrac{n_3}{N_3} + \dfrac{n_4}{N_4} + \dfrac{n_5}{N_5} + \ldots + \text{etc.}}{c}$$

where c is the number of age classes included in the calculation. All females must be aged before an unweighted mean can be calculated.

When a mean fecundity is intended for use in population analysis, the weighted form is the appropriate statistic. The unweighted mean is more appropriate to studies in reproductive physiology, and in population analysis when the age distribution of the sampled females is very different from that of the population. Sometimes the choice does not matter. The numerical difference between weighted and unweighted means, calculated from the same data, depends on the extent to which fecundity rates approaches constancy over the sampled range of ages. The weighted and unweighted means calculated from three years of age onwards are, respectively, 0·437 and 0·442 for the New Zealand red deer example (Table 7.4), a difference so small as to be trivial.

Chapter 8

Mortality

8.1 THE LIFE TABLE

The mortality pattern of a population is described by an abstract but useful construct: age-specific mortality rates are presented as if they were progressively depleting a large number of animals born simultaneously. Such a group is called a cohort. Although real cohorts are seldom studied, mortality rates calculated indirectly are applied to imaginary cohorts.

Table 8.1. Life table for female thar (Caughley 1966)

Age	Frequency	Survival	Mortality	Mortality rate	Survival rate
x	f_x	l_x	d_x	q_x	p_x
0	205	1·000	0·533	0·533	0·467
1	96	0·467	0·006	0·013	0·987
2	94	0·461	0·028	0·061	0·939
3	89	0·433	0·046	0·106	0·894
4	79	0·387	0·056	0·145	0·855
5	68	0·331	0·062	0·187	0·813
6	55	0·269	0·060	0·223	0·777
7	43	0·209	0·054	0·258	0·742
8	32	0·155	0·046	0·297	0·703
9	22	0·109	0·036	0·330	0·670
10	15	0·073	0·026	0·356	0·644
11	10	0·047	0·018	0·382	0·618
12	6	0·029			
> 12	11				

A cohort's mortality pattern is formally presented as a life table. Table 8.1 is a life table for a population of Himalayan thar. The animals are real enough, but the cohort is imaginary because the animals were not born contemporaneously. However, the table is constructed as if 205 females were born on the same day, and the numbers surviving to each subsequent birthday were recorded.

The table consists of six columns:

Column 1: age (x) at intervals of one year.

Column 2: the number still surviving (f_x) at one, two, three and so on years after birth, out of the original 205 born.

Column 3: a scaled down version of column 2 formed by dividing each value by the original strength of the cohort (i.e. 205) to give the proportion of the cohort still surviving at a given age. Each of these values is designated l_x—the probability at birth of surviving to the exact age x—and termed 'survivorship' or simply 'survival'.

Column 4: the probability of dying during the age interval x, $x + 1$. This is d_x, the frequency of mortality, calculated as the differences between two consecutive values of l_x. Although d_x refers to deaths between ages x and $x + 1$, and should logically be tabled half-way between these two ages, by convention it is tabled against the age at the beginning of the interval. Because column 3 has an initial value of 1, the d_x schedule totals 1 by definition.

Column 5: the mortality rates q_x, each of which is the proportion of animals alive at age x that die before age $x + 1$. They are calculated as d_x/l_x.

Column 6: the survival rates p_x, being the proportion of animals alive at age x that survive to age $x + 1$. They are the complements of their respective q_x values.

Life tables are not always drawn up in this way. Sometimes the l_x and d_x schedules record the history of a cohort, not in probabilities as in Table 8.1, but as numbers surviving to each age from a cohort starting with 1000, 10,000 or 100,000 animals, the first cohort strength being most common. Although the fortunes of 1000 animals are easier to visualize than a set of age-specific probabilities, only the probabilities are used in analysis. There is little point in drawing up a life table as the history of 1000 animals when each entry must be reduced to a probability before analysis can begin; and if the table is not for use in analysis there is no point in constructing it.

Life tables were first devised for human populations (apparently the first attempt was made by John Graunt in 1667). Accurate human life tables were originally for calculating life insurance premiums. They therefore include a column not provided in Table 8.1: the expectation of further life at each age. These statistics are the basic information required by the insurance industry, but they add little to an understanding of population processes. Because the methods of actuarial research on man have been taken over by biologists and sometimes applied uncritically to non-human populations, life tables for other animals often include a column of life expectancies. With the arguable exception of e_0, the life expectancy at birth, the calculation of life expectancies at each age is a waste of effort.

Because the life table contains four columns—l_x, d_x, q_x and p_x—it should not be inferred that these contain independent information. They present the same information in four different ways to allow viewing of the mortality pattern from different angles. Conversion of one column to another is no more than an arithmetic exercise. Table 8.2 expresses the algebraic interrelationships of l_x, d_x, q_x and p_x, and Figure 8.1 illustrates the relationships graphically.

Table 8.2. Interrelationship of life table statistics

In terms of

	l_x	d_x	q_x	p_x
l_x	l_x	$l_x - l_{x+1}$	$1 - (l_{x+1}/l_x)$	l_{x+1}/l_x
d_x	$\displaystyle\sum_{y=x}^{\infty} d_y$	d_x	$\displaystyle d_x \Big/ \sum_{y=x}^{\infty} d_y$	$1 - \left(d_x \Big/ \displaystyle\sum_{y=x}^{\infty} d_y \right)$
q_x	$\displaystyle\prod_{y=0}^{x-1} (1 - q_y)$	$\displaystyle q_x \prod_{y=0}^{x-1} (1 - q_y)$	q_x	$1 - q_x$
p_x	$\displaystyle\prod_{y=0}^{x-1} p_y$	$\displaystyle (1 - p_x) \prod_{y=0}^{x-1} p_y$	$1 - p_x$	p_x

$\sum$ = summation, $\prod$ = multiplication

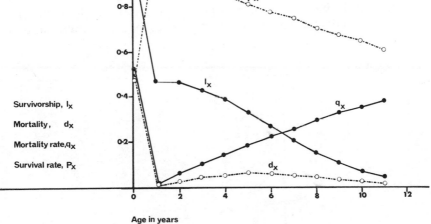

Survivorship, l_x

Mortality, d_x

Mortality rate, q_x

Survival rate, p_x

Age in years

Figure 8.1. Graphic representation of the four life table schedules of female Himalayan thar (Caughley 1966).

The four schedules of the life table are used for different purposes: l_x appears in almost all equations of population dynamics; the d_x schedule is particularly useful in studying the genetic and evolutionary consequences of mortality patterns; q_x is least affected by sampling bias, gives the most direct projection of the mortality pattern, and is the best schedule for comparisons within and between species; and p_x is used in harvesting calculations and the mathematical simulation of populations in a computer.

8.1.1 Effect of age distribution

The next chapter discusses in detail the determinants of the age distribution, but a few introductory comments are needed here because age distributions are often used to calculate life tables.

Two kinds of age distribution must not be confused: the temporal age distribution of a cohort gives the proportion surviving to each age; the standing age distribution of a population gives the number of animals, relative to the number of newborn, in each age class at a particular time. In only one case are these two age distributions congruent. When a population's exponential rate of increase is zero, and when it has been zero for some time, and when the survival and fecundity schedules have remained constant for some time, the standing age distribution is the same as the temporal age distribution of the cohorts that collectively constitute the population. In all other cases the standing age distribution differs from the cohort age distribution.

Figure 8.2 illustrates this point with an imaginary example. It presents a population that is doubling in size each year and whose mortality rate is constant with age. Consequently, decline in the size of a cohort will be linear with time when numbers are expressed as logs. The standing age distribution at any time will consist of the animals born at the most recent birth pulse, the one-year-olds of the cohort born at the previous birth pulse, the two-year olds of the cohort born at the birth pulse before that, and so on. Hence the standing age distribution must differ from any one of the cohort age distributions unless the

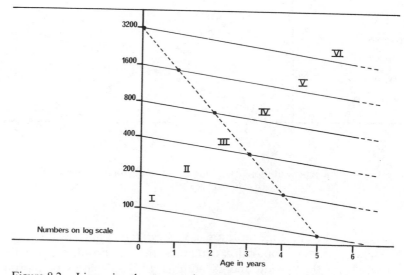

Figure 8.2. Lines give the temporal age distributions of sequential cohorts (Roman numerals) constituting an imaginary population doubling each year. The dashed line is the population's standing age distribution when cohort VI is born.

same number is born at each birth pulse, which is another way of saying that $r = 0$.

The standing age distribution in Figure 8.2 is graphed as numbers by age, but an age distribution is more conveniently expressed as the number of animals in an age class relative to the number of newborn. This age distribution, S_x, is defined as

$$S_x = F_x/F_0$$

where F_x is the number of animals aged x in the population and F_0 is the number of newborn. S_x is usually estimated from a sample by

$$S_x = f_x/f_0,$$

f_x and f_0 being sampled frequencies.

When a population's rate of increase has been constant for some time, S_x is related to survival by

$$l_x = S_x e^{rx}$$

and

$$S_x = l_x e^{-rx}.$$

Table 8.3 illustrates the dependence of age distribution on rate of increase. It lists the standing age distributions of five imaginary populations that have a common l_x schedule but different rates of increase.

Table 8.3 was constructed by calculating stable age distributions from the relationship between the l_x schedule and the rate of increase, r. The reverse procedure is also possible. A life table can be calculated from a stable age distribution if r is known. The relationship between S_x, l_x and r holds true only when survival and rate of increase have been constant for long enough to to allow the age distribution to converge to the stable form appropriate to the prevailing regime of survival and rate of increase. Irrespective of a population's rate of increase, be it positive, negative or zero, the age distribution converges towards stability and remains stable so long as l_x and r remain constant.

Table 8.3. The stable age distributions resulting from a single survival schedule interacting with different rates of increase, r

Age x	Survival l_x	Stable age distribution, S_x				
		$r = -0.4$	$r = -0.2$	$r = 0.0$	$r = 0.2$	$r = 0.4$
0	1·000	1·000	1·000	1·000	1·000	1·000
1	0·700	1·044	0·855	0·700	0·573	0·469
2	0·600	1·335	0·895	0·600	0·402	0·270
3	0·400	1·328	0·729	0·400	0·220	0·120
4	0·100	0·495	0·226	0·100	0·045	0·020

A common mistake is to assume that if an age distribution did not change during a study, the rate of increase was zero over this period.

The stable age distribution has a counterpart in the stable distribution of ages at death. Suppose all deaths occurring over a year were listed by age as a frequency distribution, and each age frequency was then divided by the total to give the proportion of deaths by age. The schedule of proportional frequencies is the distribution of ages at death, designated S'_x when it is stable. S'_x is the same as the distribution of ages at death in individual cohorts (d_x) only when $r = 0$, a point to bear in mind when constructing a life table from a picked-up sample of skulls. When the standing age distribution is stable the distribution of ages at death is also stable. It is related to the frequency of cohort mortality by

$$d_x = S'_x e^{rx}$$

or

$$S'_x = d_x e^{-rx}$$

from which it follows that $d_x = S'_x$ only when $r = 0$.

Age distributions have been divided into three categories, unstable, stable and stationary, the last being the age distribution of a population whose rate of increase is zero. Stable and stationary age distributions do not differ in kind. They are both described by

$$S_x = l_x e^{-rx}$$

which reduces to

$$S_x = l_x$$

when $r = 0$. The stationary age distribution is simply a special case of the stable age distribution—all stationary age distributions are stable but only some stable age distributions are stationary.

8.1.2 Calculating a life table

Collection of data for vertebrate life tables is usually tedious and expensive. The methods of collection and analysis should therefore be chosen carefully to give maximal efficiency. Before any data are collected, the biologist should decide whether a life table is really needed to solve the problem in hand, and if it is, how the data can be gathered with the least effort consistent with the required accuracy. Methods given below do not exhaust all possibilities but they cover the general scope of available methods.

Method 1: The number of animals dying in successive intervals of time are recorded for a group of individuals born at the same birth pulse. The frequencies are multiples of the d_x schedule and can be converted to it by dividing each frequency by the initial strength of the cohort. Table 8.4 gives the steps by which the life table is calculated. This method is applicable only to natural populations studied intensively or to captive populations.

Table 8.4. Calculation of life table statistics from the frequency of ages at death, f'_x, in a cohort

Statistic	Symbol	Calculation
Age	x	—
Number of deaths in each age interval $x, x+1$	f'_x	Original data
Probability of dying in each age interval $x, x+1$	d_x	$f'_x / \sum f'_x$
Probability at birth of surviving to age x	l_x	$1 - \sum_0^{x-1} d_x$
Mortality rate	q_x	d_x / l_x
Survival rate	p_x	$1 - q_x$

Table 8.5. Calculation of life table statistics from the number of animals in a cohort still surviving at various ages

Statistic	Symbol	Calculation
Age	x	—
Number surviving at each age x	f_x	Original data
Probability at birth of surviving to age x	l_x	f_x / f_0
Probability of dying in each age interval $x, x+1$	d_x	$l_x - l_{x+1}$
Mortality rate	q_x	d_x / l_x
Survival rate	p_x	$1 - q_x$

Method 2: The number of animals in a cohort is recorded at points of time spaced at regular intervals. These frequencies are multiples of the survival schedule, to which they are converted by dividing each by the initial strength of the cohort. Analysis proceeds as set out in Table 8.5.

Method 3: Ages at death are recorded for animals marked at birth but whose births are not coeval. The data are treated as if the animals were members of a cohort and the analysis proceeds as for method 1.

Method 4: The number of animals aged x in a standing population is compared with the number of these that subsequently die before attaining age $x+1$. The number of deaths, divided by the number alive at the beginning of the interval, provides an estimate of q_x. Human life tables are usually constructed in this way, and the method has also been used for domestic sheep (Hickey 1960, 1963). It can also be applied to a natural population. Samples taken with the same expenditure of effort at two consecutive seasons of birth would provide age distributions with numbers in each class proportional to their representation in the population at time of sampling. The number in an age class of the first sample is compared with the number in the next age class of the second sample to provide an estimate of the number dying in the year between sampling times.

The estimate of q_x for each age interval is

$$q_x = \frac{f_x \text{ (at time 0)} - f_{x+1} \text{ (at time 1)}}{f_x \text{ (at time 0)}}$$

which is actually an estimate of

$$q_x = \frac{kd_x e^{-rx}}{kl_x e^{-rx}}$$

with the constants k and e^{-rx} cancelled out. This method has seldom been used on natural population because of the difficulty of taking the same proportion of the populations at each of two samplings. A slight deviation from this constraint results in a wildly inaccurate life table. Moreover, both samples must constitute a very small proportion of the population or the first sample must be replaced after it has been recorded. However, the method has been used successfully for white-tailed deer (Eberhardt 1969a) and its application to red deer is discussed in Section 9.6.

Method 5: Ages at death are recorded for a population with a stable age distribution and known rate of increase. The data are formed into a frequency distribution of deaths in equal intervals of age, and the frequencies are each multiplied by e^{rx} to form a multiple of the d_x schedule. Subsequent calculation follows that outlined in Table 8.4. Method 5 is usually used only when rate of increase is assumed to be zero, in which case $e^{rx} = 1$ and the untreated frequencies of age at death are therefore a multiple of the d_x schedule. The method was first used by Deevey (1947) to provide a life table for the dall sheep *Ovis dalli*. Murie (1944) searched the Mt. McKinley National Park for skeletons, and determined age at death from a count of growth rings on the horns. Deevey converted these frequencies of age at death to a d_x schedule on the assumption that the population had a stable age distribution and a zero rate of increase.

Method 6: The age distribution is calculated at the birth pulse for a population with a stable age distribution and known rate of increase. The frequency of the zero-age-class is calculated from fecundity rates. Each age frequency is multiplied by e^{rx} and then divided by the number in the zero-age-class to give a table of l_x.

The data treated by this method are usually obtained by taking a sample of females at the season of births. Although the method has previously been used only for populations with an assumed zero rate of increase, it will be illustrated (Table 8.6) with data from an increasing population to make the point that r needed not be zero.

Methods 1, 2 and 3 require marking of individuals. If one must go to this trouble and expense the experiment should be designed in such a way as to

Table 8.6. Construction of a life table from a stable age distribution of female thar increasing at $r = 0.12^{a}$

Age	Sampled frequency	Correction factor	Corrected frequency	Smoothed frequency	Life table		
x	f_x	$e^{rx}(r=0.12)$	$f_x e^{rx}$	F_x	l_x	d_x	q_x
0	43^{b}	1.00	43.0	43	1.00	0.37	0.37
1	25	1.13	28.3	27	0.63	0.02	0.03
2	18	1.27	22.9	26	0.61	0.03	0.04
3	18	1.43	25.7	25	0.58	0.03	0.05
4	19	1.62	30.8	23	0.55	0.05	0.09
5	11	1.82	20.0	22	0.50	0.05	0.10
6	12	2.05	24.6	19	0.45	0.06	0.13
7	8	2.32	18.6	17	0.39	0.06	0.15
8	2	2.61	5.2	14	0.33	0.06	0.18
9	3	2.94	8.8	11	0.27	0.06	0.21
10	4	3.32	13.3	9	0.21	0.05	0.23

[a] Simplification of an analysis in Caughley (1970a).
[b] Zero frequency estimated from fecundity rates.

extract the maximum information from the data. Chapter 10 discusses how mark–recapture data can be utilized fully.

Methods 1–4 are general in that no assumption need be made on rate of increase or stability of the age distribution. In contrast, methods 5 and 6 may be used only when rate of increase is known and the age distribution is stable. In practice, a population is seldom chosen unless its rate of increase has been close to zero for some time. With the assumption that $r = 0$, $d_x = f'_x / \sum f'_x$ if method 5 is used and $l_x = f_x / f_0$ in method 6, f'_x and f_x being respectively the sampled frequency of deaths between ages x and $x + 1$, and the sampled frequency of animals aged x in the living population. Although methods 5 and 6 can be used only in a restricted range of situations they often provide the only practical means of obtaining a life table. Far too often they have been used on data that are inappropriate to this treatment. Listed below are the circumstances in which these two methods cannot be used even when the population has a zero rate of increase.

Invalid use of Method 5

1. When specimens are obtained by shooting the distribution of ages in the living population is sampled; that the animals were killed to obtain this information is irrelevant. Method 6, not method 5, is the appropriate treatment.

2. When the frequencies of ages at death are obtained only on the winter or the summer range of a population, the data reflect the mortality pattern over only part of the year and cannot be presented as a life table.

3. Where the sampled deaths resulted from a rare event such as an avalanche, flood or forest fire that removed and fixed a sample of the population's age

94

distribution during life, the resultant distribution cannot be converted to a d_x schedule. If, by chance, the catastrophe occurred at or near the birth pulse, the data could be treated by method 6 if the other necessary assumptions held.

4. When the perishability or conspicuousness of carcasses varies with age, the frequency distribution of ages at death is biased and cannot be converted to a schedule of d_x unless the bias is measured and corrected. Mortality in the first year of life is usually underestimated by a picked-up sample of skulls.

5. When some animals in a sample have died 'naturally' and the deaths of the remainder have resulted from an atypical event such as shooting or forest fire, the data contain confounded d_x and l_x effects and can be converted to neither a d_x nor an l_x schedule. There is, however, one situation in which the combining of 'natural' and hunting deaths is entirely appropriate. Where hunting is an important component of the normal hazards faced by a population, a life table combining the effects of all agents of mortality, hunting included, may be required. Hunting deaths and 'natural' deaths can then be included in the same sample, but only when the deaths resulting from these two agents are sampled in proportion to their relative effects on the population. Proportionate sampling is usually difficult and often impossible. When age at death is established at a checking station a high proportion of hunting deaths will be recorded, swamping the lower proportion of discovered natural deaths. The task is even more difficult when bands from game birds are returned from a wide area. They will represent a large number of hunting deaths but very few natural deaths.

Even in circumstances where the carcasses of shot animals are left where they fall, the hunter taking only the skin, horns or a photograph, sampling of hunting and natural deaths in their correct relative proportions is still difficult. Animals tend to be shot in open areas but they die naturally in thick cover. Deaths resulting from hunting are usually greatly oversampled in all but the most painstaking search, and the life table based on these data can be badly

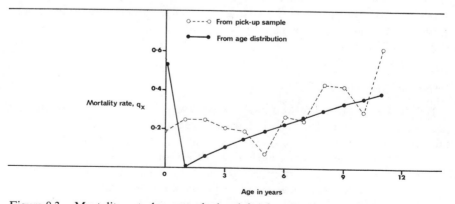

Figure 8.3. Mortality rate by age calculated for female Himalayan thar from data gathered in two different ways.

bent. Figure 8.3 demonstrates this. It shows a q_x curve constructed from a picked-up sample of 142 thar, some of which died naturally whereas the others were killed by sportsmen. It is contrasted with a more accurate q_x curve estimated from the standing age distribution at the birth pulse. The difference reflects oversampling of hunting deaths and the relatively greater perishability of juvenile skulls.

Invalid use of Method 6

1. A birth-flow population with zero rate of increase has a stable age distribution equivalent to an l_x schedule beginning from an age of half a year. The full life table cannot be calculated from the age distribution unless the rate of mortality over the first six months of life is obtained by other methods.

2. A birth-pulse population sampled mid-way between birth pulses provides the same information. The age distribution estimates survival only from six months of age onward. Likewise, a sample removed from a population by a catastrophe does not provide a full life table unless the catastrophe occurred at or near the season of births.

3. A sample shot by sportsmen, particularly when their aim is to secure meat or trophies, is likely to return a highly biased age distribution useless for calculating a life table. However, it can be used to calculate a fecundity table (Chapter 7), the accuracy of which is unaffected by a biased age distribution.

When age distributions are obtained by shooting, the sample should be collected by trained hunters capable of controlling the natural tendency to shoot the largest animals in a group. Tests for detecting biases of this kind are outlined by Caughley (1966).

Pitfalls in the path of both methods

Neither method should be used unless the assumptions on which it is based are at least approximated. Rate of increase must be known with reasonable accuracy and this rate must have remained relatively constant for two or three generations. Recent fluctuations in r rule out the use of method 6, but if the fluctuations have a periodicity much shorter than the time over which carcasses accumulated, method 5 will still return a reasonably accurate life table.

Some published life tables are based on very small samples. No absolute judgement on minimum sample size is possible because it depends on the accuracy required, which in turn depends on the use intended for the life table. But when age-distribution methods are employed, a table based on less than 150 determinations of age is unlikely to be accurate enough for any purpose.

The accuracy of a calculated age distribution depends on the size of the sample, the number of age classes, sampling bias, and the accuracy of ageing. The last is often the most important influence. Some indices of age have wide variance, and although they may be useful for dividing a sample into broad categories they cannot be used in life-table analysis. Except for ages determined by marking or banding, all methods of ageing result in some bias, and age distributions calculated by most techniques are therefore distorted to some extent. It must be left to the individual biologist to decide whether the ageing technique he uses is accurate enough to justify construction of a life table.

8.2 SMOOTHING AGE FREQUENCIES

A life table can be calculated directly from a stationary age distribution only when the frequency of each age-class x is equal to or greater than that of $x + 1$. If it were not so, one or more d_x values would be negative—a biological impossibility. When sampling variation and sampling bias violate this requirement the age distribution must be smoothed before a life table can be constructed. A smoothing formula should have two characteristics: it must be flexible enough to generate curves fitting a broad array of age distributions, and its fitting should not presuppose the relationship between mortality rate and age. Two formulae fit these requirements, the log-polynomial and the probit regression, both of which generate curves that fit most age distributions from one year of age onwards.

The log-polynomial takes the form

$$\log f_x = a + bx + cx^2 + dx^3 + \ldots \text{etc.}$$

where f_x is the sampled frequency of age x and a, b, c and d are constants. The fitting is carried out in stages, the success of fitting terms of higher and higher degree being tested after each addition. Snedecor and Cochran (1967) explain the method simply, and most computers have a standard program for this exercise. The first step calculates $\log f_x = a + bx$, a satisfactory fit implying that mortality rate is constant with age. The reduction in sum of squares due to fitting this regression is divided by the mean square of the remainder. If the ratio is significantly greater than unity the hypothesis of constant mortality rate by age is discarded and the regression

$$\log f_x = a + bx + cx^2$$

is then calculated. The process continues until the addition of further terms results in no further significant reduction in sum of squares. I have not found an age distribution whose log-polynomial, fitted from age one onwards, extended beyond the x^2 (i.e. quadratic) term.

The quadratic polynomial generates a parabola whose axis is vertical. Only the right descending limb is fitted to the logged age distribution from age one, and the axis of the parabola is usually close to this age. If, by chance, it falls to the right of age one, the curve is no longer appropriate because the fitted frequencies will rise initially before declining.

An alternative is the curve of the form

$$\text{Probit} (f_x/f_1) = a + bx^i,$$

where a, b and i are constants, fitted from age two onward. The fitting is more laborious than for the log-polynomial because the constant i must be calculated by iteration.

The steps are as follows:

1. Calculate f_x/f_1 from age two onward.
2. Transform these to probits (Finney 1947).
3. Calculate the schedule of x^i for $i = 0.1$.
4. Calculate the linear regression of probit (f_x/f_1) on x^i.
5. Convert the fitted probits from age one onwards back to fitted frequencies, calculate the squared difference between each observed frequency and its fitted frequency, and sum these.
6. Repeat steps 3, 4, and 5 for $i = 0.2$, $i = 0.3$, $i = 0.4$, and so on, to find the value of i resulting in the lowest sum of squared differences between observed and fitted frequencies.

The smoothed frequencies F_x, in Table 8.6 have been calculated in this way. It is a tedious business, but the fitting provides bonuses that compensate for the labour. The probit curves treats the underlying distribution of 'adult' (i.e. from age 1) deaths as a probability distribution. It provides insights into the mortality pattern that do not come from an inspection of the life table. The distribution of adult deaths by age is describable directly from the constants of the probit curve by median age of adult death

$$= \left(\frac{5-a}{b}\right)^{1/i}$$

and standard deviation of adult death

$$= \frac{1}{2}\left[\left(\frac{4-a}{b}\right)^{1/i} - \left(\frac{6-a}{b}\right)^{1/i}\right].$$

An index of the skew of adult ages at death is provided by $-\log_e i$. When $\log_e i = 0$ the distribution is normal, when negative the distribution is negatively skewed, and when positive the distribution is positively skewed.

8.3 PARTITIONING MORTALITY RATES

Population managers need to know the proportionate effects of several agents of mortality. Usually we wish to isolate the effect of hunting from the effect of all other influences on mortality, a group usually lumped together as the 'agents of natural mortality'. If u_x and w_x are the proportions of age class x killed over a year by hunting and by natural agents, respectively, and q_x is the proportion of the age class killed by all agents,

$$q_x = u_x + w_x.$$

Values of u_x and w_x are not independent of each other. If hunting were banned, w_x would increase because, in the absence of hunting, many animals that would otherwise have been shot are now at risk of death from other agents. So u_x and w_x are not particularly helpful in partitioning the effects of the two causes of mortality. The problem is turned by defining mortality rates as h_x, the proportion of age-class x that would die by hunting if no animals died from natural causes, and n_x, the proportion that would die from natural causes if there were no hunting. These are termed respectively the isolated rate of hunting mortality and the isolated rate of natural mortality. They are related to the total rate of mortality by

$$q_x = h_x + n_x - h_x n_x,$$

which can be expressed in terms of the equivalent survival rates as

$$p_x = (1 - h_x)(1 - n_x).$$

Ricker (1958) pointed out that in making use of this relationship there is no need to assume that the isolated rates of hunting and natural mortality are proportional throughout the year. Rates of $h_x = 0.5$ and $n_x = 0.5$ will result in a total rate of $q_x = 0.75$, whether the two agents of mortality act consecutively or concurrently. This convenient relationship does not hold for u_x and w_x, the actual rates of mortality.

Suppose we wished to partition further by dividing hunting mortality into the component effects of hunting by sportsmen and hunting by professionals. The isolated rate of mortality referrable to sportsmen will be termed g_x and that referrable to commercial hunting will be termed h_x. Then

$$p_x = (1 - g_x)(1 - h_x)(1 - n_x)$$

which multiplies out in terms of $q_x (= 1 - p_x)$ to

$$q_x = g_x + h_x + n_x - g_x h_x - g_x n_x - h_x n_x + g_x h_x n_x.$$

Analogous equations can be written for any number of agents of mortality.

8.4 SEASONAL TRENDS IN MORTALITY

A life table with age intervals of one year is adequate for most purposes. Occasionally, however, a closer look at mortality is called for, particularly when it has a marked seasonal trend. Seasonal changes in the incidence of mortality can best be measured by the number of dead animals discovered each month in a circumscribed area. These data are then reduced to monthly rates of mortality by a method applicable equally to birth-pulse and birth-flow populations. The example (Table 8.7) uses the counts of dead adult grouse reported by Jenkins and Watson (1962) from their study area in Scotland. Their figures for 1957–8 and 1958–9 are pooled, and for purposes of illustrating the analysis I have arbitrarily assumed that the annual mortality rate of adults is 40 per cent.

Table 8.7. Analysis of seasonal variation in the rate of mortality of red grouse (data from Jenkins and Watson, 1962).

Month m	Mortality frequency f'_m	Exposed to risk F_m	Mortality rate f'_m/F_m	Survival rate $1-(f'_m/F_m)$
Jul.	14	1,160	0·012	0·988
Aug.	9	1,146	0·008	0·992
Sep.	17	1,137	0·015	0·985
Oct.	17	1,120	0·015	0·985
Nov.	42	1,103	0·038	0·962
Dec.	92	1,061	0·087	0·913
Jan.	37	969	0·038	0·962
Feb.	64	932	0·069	0·931
Mar.	57	868	0·067	0·933
Apr.	46	811	0·057	0·943
May.	50	765	0·065	0·935
Jun.	19	715	0·027	0·973
	$\Sigma = 464$			$\prod = 0·600$

The analysis proceeds as follows:

1. Table the number of adults found dead per month, the table starting from the month after the birth pulse.
2. Divide the total number of dead birds found (464 in the example) by the adult mortality rate per year (0·4) to give the theoretical number of adults exposed to risk at the birth pulse. Table this value (1,160) against the first mortality frequency.
3. Subtract the mortality frequencies in sequence from the initial number exposed to risk, and table each intermediate sum.
4. Divide each mortality frequency by its appropriate 'number exposed to risk' to give mortality rate per month.
5. Table the complement of each monthly mortality rate, i.e. the monthly survival rate.
6. Check the calculation by multiplying together all monthly survival rates. The product will equal the complement of the annual mortality rate (0·4) introduced in step 2.

8.5 MORTALITY PATTERNS

Mammals

Caughley (1966) summarized the existing information on mortality patterns in mammals, concluding that mortality rate typically follows a 'U' shaped trend with age. The few mammalian life tables published subsequent to 1966 have strengthened this conclusion.

Prepubertal mortality rates tend to vary greatly between populations of one species. Man is no exception. In 1968 I estimated prepubertal mortality in the hills of Nepal as approximately 70 per cent; in Canada the current rate is approximately 3 per cent (WHO 1972). Adult mortality rates of mammals are much less variable. Compared with fluctuations in prepubertal mortality, year to year variation in the mortality rate of adults contributes little to fluctuations in rate of increase.

Birds

An excellent review of mortality patterns in birds was published by Farner (1955). It should be read in full by anyone interested in this topic.

Avian life tables are difficult to interpret for two reasons:

1. Initial work in this field by Nice (1937), Kraak *et al.* (1940) and Lack (1943a, b, c) detected a tendency towards constancy of mortality rate by age. Although the data were neither numerous nor controlled in accuracy they justified a *hypothesis* that the rate of adult mortality was constant with age. By some accident of history ornithologists did not treat the hypothesis as a question to be examined and tested. It was accepted first as a conclusion, then as a principle, and finally as an axiom. Most studies on mortality in birds now begin from the axiom that mortality rate is constant after the first year. The survey design and subsequent analysis is usually shaped by the axiom to the extent that the results no longer test the hypothesis that adult mortality rate is constant.

2. Almost all studies of mortality depend on banding, and bands can drop off. Measures against band loss differ between studies. In some, worn bands are replaced at recapture; in most, the problem is ignored and mortality rates are calculated as if no losses occurred. I do not know of a study in which loss of bands has been accurately estimated and the life table corrected for this bias.

A correction for band loss is given in Chapter 10, but the question at this stage is whether band loss distorts a life table. The problem can be investigated by a simple model in which loss of bands is initially rare but steadily increases with time after banding. If A is the rate of band loss in the first year, and this rate multiplies by e^i in each subsequent year, the rate of loss during year t after banding is $A e^{it}$. The proportion P_t of birds banded as fledglings that still retain their band after t years is

$$P_t = \prod_{y=0}^{t} 1 - A e^{iy}$$

Figure 8.4 shows the model in action. The 'apparent' mortality rate of 0·45 per year is calculated from birds found dead. The 'actual' mortality rate is calculated on the premise that the cohort suffered a band loss of 1 per cent in the first year, the rate increasing by a factor of $e^{0.3}$ per year thereafter. 'Actual'

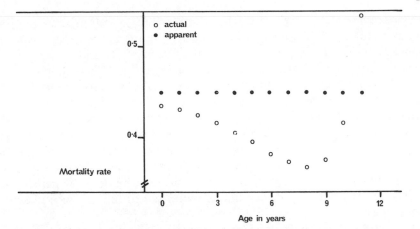

Figure 8.4.   A possible effect of band loss on the calculated trend of mortality rate with age. See text.

and 'apparent' mortality patterns are very different. This result is not an argument that an apparently constant rate of mortality is, in actuality, a 'U' shaped curve, or that the parameter values of the model are real, or even that the model itself is appropriate. These points are irrelevant. The model's purpose is to demonstrate that band loss can distort mortality statistics, and that a life table based on band recoveries, if uncorrected for loss of bands, must be viewed with mild scepticism.

Despite these difficulties a few generalizations can be made. It does appear that mortality rates change less with age than they do for mammals. Richdale's (1954) tables for yellow-eyed penguins, which were not based on recovery of bands, demonstrate this clearly. The difference between prepubertal mortality rate and that at early adulthood also appears to be smaller in birds than in mammals.

Fish

The majority of exploited populations of fish show a decline of age frequencies consistent with a constant rate of mortality from adulthood onwards. However, data from unfished or lightly fished populations (Ricker 1949, Kennedy 1953 and 1954, Wohlschlag 1954, and Tester, quoted in Ricker 1958) indicate that in these populations older fish have a mortality rate higher than middle-aged fish.

Prepubertal mortality is very great in all fish populations, but it is difficult to measure. Allen's (1951) estimate of 99 per cent first-year mortality in brown trout probably indicates the general order of juvenile mortality in fish.

Amphibians and reptiles

There seems to be no complete life table published for any frog species.

The closest approach is provided by age distributions of postmetamorphic anurans. However the most comprehensive, those for *Bombina bombina* (Bannikov 1950), *Pseudacris brachyphona* (Green 1957) and *Rana pretiosa* (Turner 1960a, 1962), transmit no common message. When graphed as a survival curve the age distribution of *Bombina* adults suggests a mortality rate declining with age, that for *Pseudacris* suggests that the rate of mortality is independent of age, and the trend for *Rana* differs markedly between successive years. These apparent differences do not necessarily imply large differences between the mortality patterns of anuran genera. Ageing inaccuracy and variation in rate of increase are more likely explanations. In each case age distributions were translated directly from size distributions, a method that bends survival curves by an amount proportional to the variance of size at a given age.

Organ (1961) calculated survival curves from age distributions for five species of the salamander genus *Desmognathus*. Replicates varied considerably within species but the pooled data are consistent with a rate of mortality increasing with age.

Until very recently there were no complete life tables for reptiles. Over the last few years however, probably under the stimulus of the pioneering work of D. W. Tinkle, herpetologists have been producing life tables at a rapid rate. These suggest that reptiles have a mortality pattern similar to that of mammals: a high rate of mortality over the first year, a lower rate in the second year and a progressively increasing rate thereafter. As a typical example, a population of *Sceloporus graciosus* had mortality rates over the first four years of life of 0·77, 0·38, 0·47 and 0·55 respectively (Tinkle 1973). Most lizards die young. Tinkle (1967) showed for *Uta stansburiana* that fully a third of all first-year deaths occurred in the first week.

8.6 MISINTERPRETATION OF STATISTICS

Potential longevity

Throughout most writing on the ecology of vertebrate populations runs an undercurrent view of mortality. Death is treated as an event that should decently occur at a certain pre-ordained age after a full life; death before this age is, in the medieval sense, an accident. From this view of death derives the concept of 'potential longevity', an age that all members of a species would inevitably reach were it not for environmental restraints (see particularly Bodenheimer 1958). The mortality pattern of a population is thereby considered a product of two influences: the animal's intrinsic inability to remain alive after a certain age, and 'environmental resistance' that snatches away many lives before 'potential longevity' is realized.

The lifespan of an individual animal is unambiguously the interval between its birth and its death. Equally precise is the definition of the mean lifespan of individuals in a population, i.e. the life expectancy at birth. But these statistics are not being sought by the question 'how long can an elephant live'? The questioner is asking for an estimate of 'potential longevity' and must be told that there is no such age. A questioning of the reality of 'potential longevity' is often met by some variation on the counter-question, 'do you believe that an animal can live for 1000 years'? This question is irrelevant because it side-steps the meaning of 'potential longevity'. It is an age to which animals of a given species can live in exceptional conditions, but beyond which they cannot live. Because animals do not live to an age of 1000 years, or to infinity, there is no justification for claiming a bounding age less than this.

If there were such an age the discovery of this fact would be simple. It would show up as an inflection in the plot of mortality rate against age, particularly in captive populations living in near-optimal conditions. The most accurate life tables available for man and captive mammals provide not the slightest evidence of either a strict potential longevity or a mean potential longevity surrounded by a variance.

'Potential longevity' is not an appropriate general concept. It must be replaced by a probability statement. An animal dies before the age of infinity not because it cannot pass some bounding age but because the probability of its riding out the ever present risk of death for that long is infinitesimal. This statement is not advanced as convenient mathematical sophistry. It is likely to be a closer approximation to biological reality than a model utilizing a cut-off. Cut-off effects are rare in biological systems and no evidence dictates their necessity in a model of mortality.

Life expectancy

Life expectancy is the mean age at death of the members of a cohort. Although unambiguous it is commonly misinterpreted.

Life expectancy in Mabuta is 30 years. This does not imply that most Mabuta-nese die between 25 and 35 years of age; nor does it imply that few old men live in Mabuta; nor does it imply that a Mabutanese aged 50 years is physio-logically older than his contemporary in California: it tells us only that juvenile mortality is high in Mabuta. If the juvenile mortality is lowered by public health measures, the life expectancy will rise. The rise is an arithmetic conse-quence of lowered juvenile mortality, occurring whether or not adult mortality rates are also lowered.

The fallacies surrounding 'potential longevity' and 'life expectancy' can teach us something. The difficulties produced by the first indicate that the terms by which we describe a mortality pattern must be defined rigorously. Otherwise we drift into metaphysics. The second shows that the description of a mortality pattern by a single statistic can lead to a murky appreciation of what is actually happening to the population.

8.7 SUMMARIZING A LIFE TABLE

A full life table is often not necessary to solve a problem, and more often a life table, although fervently desired, is not obtainable for practical reasons. Sometimes the initial construction of a detailed life table indicates that further study can be restricted to selected intervals of age. In each of these cases a contracted life table can be constructed, the degree of contraction depending on the table's intended use. Life tables for people are usually pooled by periods of five years, and a similar pooling may be justified for other species with low rates of adult mortality. But more often the life table is integrated over uneven intervals of age, each combining a set of ages that share the same characteristics. A common division is into the intervals 'juvenile', 'yearling' and 'adult', or even 'juvenile' and 'post-juvenile'. Mortality over each interval is given as a mean rate per year of age. The mortality rate for juveniles and for yearlings can be read off the life table as q_0 and q_1 respectively, but that for adults must be calculated as

$$\bar{q}_{2,\infty} = \frac{d_2 + d_3 + d_4 \ldots + d_\infty}{l_2 + l_3 + l_4 \ldots + l_\infty}$$

$$= l_2 \bigg/ \sum_{x=2}^{\infty} l_x :$$

$\bar{q}_{2,\infty}$ is the weighted mean mortality rate per year for animals aged two years onward. If the adults must be further divided into 'young' adults (say ages 2 to 10) and 'old' adults (age 10 onwards) the mortality rates would be

$$\bar{q}_{2,10} = \frac{d_2 + d_3 + d_4 + \ldots d_9}{l_2 + l_3 + l_4 + \ldots l_9}$$

$$= \frac{l_2 - l_{10}}{\sum_{x=2}^{9} l_x},$$

and

$$\bar{q}_{10,\infty} = l_{10} \bigg/ \sum_{x=10}^{\infty} l_x.$$

The weighted mean mortality rate for the entire population gives the proportion of animals alive at one birth pulse that have died before the next:

$$\bar{q} = \sum d_x e^{-rx} / \sum l_x e^{-rx}$$

which simplifies to

$$\bar{q} = 1 / \sum l_x$$

when the population is stationary. The latter equation is a mathematical identity of the more cumbersome formula used in ornithology, where $\bar{q}$, usually termed M_w, is calculated from the number of cohort deaths D in the 1st, 2nd, 3rd, etc. interval of age, by

$$M_w = \frac{D_1 + D_2 + D_3 + D_4 + \text{etc.}}{D_1 + 2D_2 + 3D_3 + 4D_4 + \text{etc.}}.$$

The population mortality rate $\bar{q}$ is a more precise measure of mortality rate than is 'turnover time'. The latter measures the time needed for natural replacement of all individuals in the population. 'Turnover time' and its modifications '95 per cent turnover time' and '90 per cent turnover time', are commonly used in wildlife research. However, they share few of the advantages of $\bar{q}$ and can claim several unique disadvantages.

Ratio methods

A large number of formulae are available (Kelker 1949–50, 1952, Robinette 1949, Petrides 1954, Selleck and Hart 1957, Hanson 1963, Kelker and Hanson 1964) to estimate rates of hunting mortality and total mortality for various segments of the population. They operate on proportions of juveniles, yearlings and adults in a sample. Many of these methods provide a mortality rate for óne segment expressed as a ratio of that of a second segment, rather than absolute rates of mortality. These methods tend to provide answers irrelevant to most practical or theoretical problems.

Most age-ratio methods are based on the assumption, often undeclared, that the population has a stable age distribution appropriate to a zero rate of increase. They cannot therefore be used to track changes in mortality rate from year to year. Of the various age-ratio methods available, only two are recommended as useful in estimating mortality rates, and then only in special circumstances. One of these (Kelker's method) was discussed in Section 4.2.4 because its estimate of mortality is secondary to its use in estimating density. The other provides a rough estimate of $\bar{q}$ for an approximately stationary population. When j is the number of newborn in a random sample of size n taken at the birth pulse,

$$\bar{q} = \frac{j}{n}.$$

Life expectancy at birth is therefore

$$e_0 = \frac{2n - j}{2j}.$$

When these formulae are applied to data from a birth-flow population, $\bar{q}$ estimates the mortality rate per year from six months of age and e_0 estimates life expectancy from a base age of six months. Derivations and examples are given by Caughley (1967b).

8.8 THE USES OF A LIFE TABLE

To the ecologist, the main value of a life table lies in what it tells him about a population's strategy for survival. He needs to know the demographic mechanisms by which, over the long run, rates of birth and death are equalized to hold the mean rate of increase at zero. Differing attacks on this problem have been made by Cole (1954), Lewontin (1965), Murdoch (1966), Holgate (1967), Frank (1968), Murphy (1968), Reddingius and den Boer (1970) and Meats (1971).

The ecologist also needs to know how the population's pattern of mortality changes over the short run in response to environmental fluctuations. Paradoxically, if he is trying either to reduce a population or to stimulate its increase, he must identify the weakest link in the life history. This stage is attacked if the aim is control or pampered if the goal is conservation. The Achilles' heel of a population is usually that age class showing the greatest between-year variability in either mortality rate or fecundity rate, or in the interaction between the two. It can be identified only by comparing mortality and fecundity schedules from several years. A persuasive demonstration of the method's utility is provided by the Canadian research on spruce budworm (Morris 1957, 1963).

A less empirical variant of this approach is the prediction of a population's reaction to a given set of circumstances by modelling the population's dynamics. The trend of numbers on time is calculated (usually by computer) for different initial age distributions (e.g. Niven 1970), or for various modifications, either fluctuating or constant with time, to the schedules of fecundity and survival (e.g. Lefkovitch 1967, MacArthur 1968). Such simulations are particularly useful for estimating the effects of age-selective harvesting (Section 11.2). Modelling has many applications, not the least of which is the speed with which it identifies and rejects false hypotheses. Quite simple simulations are usually sufficient to eliminate those hypotheses that are certainly untrue, leaving a small number than can often be tested by a single carefully designed experiment.

An estimate of intrinsic rate of increase, that rate at which a population increases when no resource is limiting, is often needed to answer questions of ecology (Birch 1948, Nagel and Pimentel 1964), genetics (Fisher 1930; Hairston, Tinkle and Wilbur 1970), evolution (Birch et al. 1963) and sustained-yield harvesting (Beverton and Holt 1957, Ricker 1958, and Section 11.2). The most accurate estimates are calculated from a life table together with a fecundity table (Section 9.2).

Chapter 9

Relationship between parameters

9.1 THE BASIC EQUATION

The schedules of survival and fecundity are related to rate of increase by Lotka's (1907, a and b) equation

$$\sum l_x e^{-rx} m_x = 1.$$

A little time given to understanding this equation is well spent. It is the basic equation of population dynamics and most statistics of population analysis are derived from it. It expresses the dynamics of the female segment of a population having fixed schedules of survival and fecundity, i.e. the rates of survival and fecundity may differ by age but the schedules taken as a whole do not change with time.

This is an abstraction from nature because survival and fecundity schedules seldom remain constant for long. Yet the complexities of natural populations cannot be understood unless they are studied in terms of the basic equation and described in terms of their deviation from the simple system that the equation represents.

The equation's logic will be explored through an imaginary example. At the birth pulse occurring at time zero one female is born. The population is increasing at the rate r with constant schedules of fecundity and survival. The number of females born at the previous birth pulse will therefore have been e^{-r}. Of these, $l_1 e^{-r}$ will have survived one year of life to be present in the population at $t = 0$, and they will contribute $l_1 e^{-r} m_1$ female births to the $t = 0$ birth pulse. A similar calculation can be made for each age. For instance, at $t = -3$, $e^{-r} \times e^{-r} \times e^{-r} = e^{-r3}$ females were born, $l_3 e^{-r3}$ of these survived to become three-year-olds in the population at $t = 0$, and they contributed $l_3 e^{-r3} m_3$ female births to that birth pulse. (If the reader has difficulty conceptualizing fractional animals he can start with 1000 females rather than one female at time zero.)

Table 9.1 gives these statistics for a population of five age classes. The last column lists the number of female births contributed to the $t = 0$ birth pulse by each age class, and the sum of these subtotals must equal the total number of

Table 9.1. Contribution of different age classes to the number of females born at birth pulse zero

Time	Females born	Their ages at time zero	Number surviving at time zero	Number of female births produced at time zero
0	1	0	1	0
-1	e^{-r}	1	$l_1 e^{-r}$	$l_1 e^{-r} m_1$
-2	e^{-r2}	2	$l_2 e^{-r2}$	$l_2 e^{-r2} m_2$
-3	e^{-r3}	3	$l_3 e^{-r3}$	$l_3 e^{-r3} m_3$
-4	e^{-r4}	4	$l_4 e^{-r4}$	$l_4 e^{-r4} m_4$
-5	e^{-r5}	5	$l_5 e^{-r5}$	$l_5 e^{-r5} m_5$

$$\text{Total Births} = \sum l_x e^{-rx} m_x = 1.$$

females born at $t = 0$. But this total was set initially at one, and therefore the sum of the values in the last column of Table 9.1 equals one by definition. Hence

$$\sum l_x e^{-rx} m_x = 1.$$

The variables in this equation have been defined in terms of the female segment of the population. With equal validity they could have been expressed as male survival and male births per male. So long as both female and male schedules of survival and fecundity are fixed the value of r appropriate to the male segment is the same as r for the female segment. The principle holds whether or not survival and fecundity schedules are the same for males and females.

The basic equation need not be defined in terms of survival from birth and of births per female. Birth is an arbitrary point in the life span and any other arbitrary point is equally valid as a demarcation of fecundity from survival. Up to this point the existence of an individual is ignored in analysis, at this point it becomes a component of fecundity, and beyond this point it becomes a survivor. If fecundity were defined as female conceptions per female, survival would be defined as the probability at conception of surviving to various ages. Or fecundity could be defined as the production of female yearlings per female, and survival would then be the probability at one year after birth of surviving a further one year, two years, three years, and so on. But one thing cannot be done. If fecundity and survival are determined from different base ages—say fecundity defined as frequency of conceptions but survival as the probability of surviving from birth—the basic equation cannot be used to express the relationship between survival, fecundity and rate of increase.

9.2 CALCULATION OF r

Chapter 5 introduced several concepts of rate of increase. They differ less in the techniques by which they are estimated than in their interpretation:

$\bar{r}$ = observed rate of increase.

r_s = survival-fecundity rate of increase, the rate implied by the prevailing schedules of survival and fecundity.

r_m = intrinsic rate of increase, the rate achieved in the absence of crowding and of shortage of resources.

ρ = the intrinsic rate of increase in the best of all possible environments (Cole 1957).

r_p = potential rate of increase, the rate that would result if the effect of a given agent of mortality were eliminated.

Each of these rates is designated by the generic symbol r. Subscripts are used only when a particular interpretation of r must be specified.

Suppose I were managing a population of deer for sustained yield. I would need to know how much the population fluctuated from year to year. This would be measured by several yearly estimates of $\bar{r}$. I might also need to know the extent to which the fluctuations in $\bar{r}$ were determined by fluctuations in survival and fecundity, as against changes in age distribution. Estimates of r_s would reveal the rates of increase implied by the prevailing survival and fecundity, the effect of unbalanced age distribution being eliminated. The appropriate rate of harvesting would be determined by an estimate of r_p, the rate of increase that would result if all hunting were stopped. It is identical with the rate of harvesting that holds the population to a zero mean rate of increase. Should I wish to maximize the harvest I would need an estimate of the maximum rate at which the population would increase in this environment if the effects of crowding and of shortage of resources were completely eliminated. This rate is r_m. It is needed to calculate the maximum sustained yield (see Chapter 11).

Calculating r from population estimates

The observed exponential rate of increase, $\bar{r}$, can be calculated by regression analysis from two or more estimates of population size. If the population is of the birth-pulse type the estimates must be made at about the same date each year. Estimates, or indices of the estimates, are first converted to natural logs. These will each be designated N. Time units of one year, preferably scaled so that the first equals one, will each be designated t. Then the mean exponential rate of increase $\bar{r}$ over the period of observation is

$$\bar{r} = \frac{\sum Nt - (\sum N)(\sum t)/n}{\sum t^2 - (\sum t)^2/n}$$

where n is the number of estimates.

Calculating r from survival and fecundity

The regression method calculates a value of r averaged over a period of time. Estimates of r_m and r_s are instantaneous measurements that must be calculated from survival and fecundity schedules by way of the basic equation

$$\sum l_x e^{-rx} m_x = 1;$$

but this equation cannot be solved directly for r except by complex methods beyond the scope of this book. Instead, trial values of r can be fed into the LHS to satisfy RHS $= 1$. The labour of this iterative process should be left to a computer. Appendix 2 gives a FORTRAN program that will estimate r in a few seconds.

A few tricks speed the calculation if it is to be run on a desk calculator. The steps are:

1. Draw up a column of the $l_x m_x$ values.
2. Calculate r_a, an approximation to r, by

$$r_a = \frac{\sum l_x m_x \log_e \sum l_x m_x}{\sum l_x m_x x}$$

3. Table the values of $r_a x$ for each age, and next to them the appropriate values of $e^{-r_a x}$ which can be read off the table in Appendix 1.

4. Now table a column of $l_x m_x e^{-r_a x}$ and sum them. The total usually will be greater than 1, indicating that r_a is less than r.

5. Select a second trial value of r by multiplying r_a by about 1·3. This value will be designated r_b. Repeat steps 3 and 4 with r_b to provide a second summation. It will be less than 1 if r_b is greater than r. If so, r has been bracketed between r_a and r_b. If not, increase r_b until the summation is less than 1.

6. Graph r_a and r_b against their respective solutions of $\sum l_x e^{-rx} m_x$ and link the points with a straight line. The value of r on this line corresponding to $\sum l_x e^{-rx} m_x = 1$ is a close approximation to the true value of r. Although it is close enough for practical purposes it is a slight overestimate because the regression of trial values of r on the solution of $\sum l_x e^{-rx} m_x$ is gently curved.

7. When greater accuracy is required, the interpolated estimate of r is fed into the basic equation to provide a summation, and a second summation is generated with a slightly lower trial r. A more accurate interpolation of r is then estimated graphically.

Calculating r by matrix algebra

An alternative and direct method calls for a Leslie–Lewis matrix of survival and fecundity rates, r being solved as the natural logarithm of its dominant latent root (Lewis 1942, Leslie 1948). Several recent ecological texts extolling the elegance of this method ignore its severe practical limitations. The Leslie–

Lewis rates of survival (P_x) and fecundity (F_x) are different measures from the p_x and m_x used here, F_x in particular being difficult to calculate from field data. It is defined (Leslie 1948) as 'the number of daughters born in the interval t to $t + 1$ per female alive aged x to $x + 1$ at time t, *who will be alive in the age group 0 to 1 at time t + 1.*' The emphasis is mine; the problem of how to estimate this biologically complex statistic in the field is yours. Formally, $F = e^r m_x$ (Pielou 1974), but that relationship does not normally provide a short-cut to estimating F_x because r, far from being a statistic calculated independently, is the unknown we seek from independently estimated fecundity rates and survival rates.

Because the operations of matrix algebra parallel the operations by which a computer thinks most efficiently, matrices are convenient tools for modelling the growth of populations. They are less useful at the tactical level of analysing the dynamics of a natural population because they are difficult to construct from field data.

Calculating r from survival and age distribution

When a population's age distribution is stable, S_x formed by dividing the number of females of age x by the number of newborn females is related to survival by

$$S_x = l_x e^{-rx}$$

Consequently r can be calculated for a birth-pulse population if l_x and S_x are known for a single age:

$$r = \frac{\log_e l_x - \log_e S_x}{x}$$

A more accurate estimate is the unweighted mean of r calculated in this way for several age classes.

9.3 MAXIMUM RATE OF INCREASE

The maximum rate at which a population can increase in a given environment is r_m for that environment. In the best of all possible environments r_m reaches a maximum of ρ. Both of these rates are defined by r in

$$\sum l_x e^{-rx} m_x = 1$$

where r can be r_s, r_m or ρ according to the conditions under which the l_x and m_x schedules were measured. These rates of increase are resultants of fecundity interacting with mortality.

But there is another rate, the 'maximum rate of increase', that is often mentioned but seldom defined. It lurks behind such statements as 'rabbits, if uncontrolled, will cover the surface of the earth in 8.3 years', and 'one pair of chipmunks in ideal conditions will give rise to 363,271,486 descendants in five years'.

Table 9.2. Estimation of intrinsic rate of increase, r_m, from selected populations

Species	r_m/year	Conditions	Data from:
Short-tailed vole	4·56	laboratory	Leslie and Ranson 1940
Norway rat	3·91	laboratory	Leslie *et al.* 1952
Orkney vole	3·74	laboratory	Leslie *et al.* 1955
Whitetailed deer	0·55	field	Kelker 1947
Pacific sardine	0·39	field	Murphy 1967
Wapiti	0·27	field	Murphy 1963
Roe deer	0·23	field	Andersen 1962
Man	0·04	field	Various

Although these estimates are usually regarded as unrealistic in that the necessary ideal conditions never occur, seldom are they recognized as nonsensical. They express the result of a rate of increase generated by maximum fecundity in the absence of deaths. For good psychological reasons 'maximum rate of decrease' is never defined as r resulting from absence of births, but if one is accepted the other logically follows.

This concept of 'maximum rate of increase' may be quaint but it is certainly not harmless. It has been used to justify control measures of doubtful value, and ecological studies stressing reproduction while ignoring mortality. This is a convenient dereliction since fecundity is much easier to measure than is mortality, a difference accurately reflected in the relative lengths of Chapters 7 and 8. Reproduction is often thought of as the basic population process, to which mortality is subordinated. But dying is a biological process no more nor less real than reproduction. An understanding of a population's dynamics rests on an understanding and measurement of both. Fecundity alone tells nothing about a population's maximum or actual rate of increase; mortality alone is equally reticent. The maximum rate of increase in a given environment is r_m measured from both fecundity and mortality. In few cases will it exceed 0·4 per year for large mammals and 4·0 for small mammals and birds. Table 9.2 shows selected estimates of r_m.

9.4 STABILITY OF PARAMETAL RELATIONSHIPS

When schedules of survival and fecundity are fixed, the age distributions of males and females each converge to a stable form and the rate of increase becomes constant (Lotka 1907a and b; Sharpe and Lotka 1911). The effect occurs whether rates of mortality are constant with age or not, whether the survival and fecundity schedules of males and females are the same or different, and whether the rate of increase intrinsic to these schedules is positive, negative or zero. The principle carries two corollaries:

1. Each segment of the population (e.g. males, females, year classes, breeders,

non-breeders, juveniles and adults, and any combination or subdivision of these) increases at the same rate r when the stable rate of increase is achieved.

2. If any one segment of the population, excepting a post-reproductive stage, shifts to a new rate of increase and holds to this rate, the rate of increase of all other segments will converge to it.

Figure 9.1 diagrams the relationship between four segments of the population. A white arrow indicates that the rate of increase of one strongly influences the rate of increase of the other. A black arrow identifies a weaker effect that requires more time to equalize the two rates of increase.

These relationships are important in wildlife management because they permit great flexibility in manipulating the size and rate of increase of a population. If, for example, an aim of management were to stabilize the numbers of a growing population, the result could be achieved by stabilizing numbers in a single segment. The size of all other segments would then stabilize in sympathy because their rates of increase must converge to that ($r = 0$) of the manipulated segment. The ability to manipulate an entire population by acting on only one or two segments is of practical importance when some segments are cryptic, nocturnal or dangerous. A tracing of the white arrows in Figure 9.1 will indicate that rapid stabilization results from holding either pre-reproductive or reproductive females to a rate of increase of zero. Over

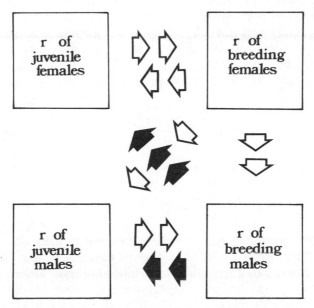

Figure 9.1. The dependence of rate of increase, r, of one segment of the population on the rate of increase of the others. White arrows indicate strong effects and black arrows weaker effects.

a much longer period the same result is achieved by stabilizing one or other of the male segments.

9.5 CONSTANCY OF SEGMENTAL PROPORTIONS

When schedules of survival and fecundity are fixed and the population has achieved its stable rate of increase, the proportion of the population held in each segment is also fixed and therefore the ratio of numbers in any two segments is fixed. It remains so while survival and fecundity schedules remain constant. A consequence of this effect—the stable age distribution—has already been mentioned. To this can be added a stable sex ratio, a stable ratio of breeders to non-breeders, and a stable proportion of newborn animals entering the population each year.

Stable age distribution

When the rate of increase has converged to a constant which depends on the fixed schedules of survival and fecundity, the relative frequencies of each age class are locked into a fixed relationship with survival. For a birth-pulse population

$$f_x/f_0 = l_x e^{-rx}$$

where f_x is the number of animals aged x at the birth pulse and f_0 is the number of newborn in the population at the same time. This ratio is the stable age frequency S_x, and a schedule of S_x comprising all ages is the stable age distribution. Males and females have different stable age distributions unless their survival schedules are identical.

The stable age distribution of a birth-flow population is related to survival and rate of increase by

$$\frac{f_{x,x+1}}{f_{0,1}} = \frac{L_x e^{-rx}}{L_0}$$

where L_x is the mean survival between ages x and $x + 1$. It is approximated by

$$L_x = \frac{l_x + l_{x+1}}{2}$$

Age classes of a birth-flow population are expressed as intervals of age. Hence $f_{x,x+1}$ is the number of animals at a specific time whose ages fall between x $x + 1$. The stable age distribution of a birth-flow population is the same throughout the year. That of a birth-pulse population changes seasonally, but it is the same when measured at the same date in consecutive years.

Although the stable age distribution is defined only in terms of r and l_x, it is also influenced by m_x through the dependence of r on fecundity. A change in the rate of survival or fecundity at a single age will dislocate the prevailing

Table 9.3.   The effect on age distribution of an 'across the board' increase
in mortality

Age x	Original		Modified	
	Survival l_x	Age distribution S_x	Survival l'_x	Age distribution S_x
0	1·000	1·000	1·000	1·000
1	0·845	0·692	0·761	0·692
2	0·824	0·552	0·667	0·552
3	0·795	0·436	0·579	0·436
4	0·755	0·339	0·495	0·339
5	0·699	0·257	0·413	0·257
6	0·626	0·189	0·333	0·189
7	0·532	0·131	0·255	0·131
8	0·418	0·084	0·180	0·084
9	0·289	0·048	0·112	0·048
10	0·162	0·022	0·057	0·022
11	0·060	0·007	0·019	0·007
	$r = 0·2000$		$r' = 0·0945$	

stable age distribution. The age distribution will then converge to another
stable form appropriate to the new regime of survival and fecundity. When
these schedules fluctuate continually the age distribution lags several steps
behind, hunting for a stable form appropriate to a fecundity-survival regime
that has already changed.

Leslie (1948) showed that an 'across the board' change in mortality rate
affecting all age classes equally does not change the age distribution. The survival
schedule changes, as does the rate of increase, but these two changes exactly
compensate to hold the age distribution constant.

Table 9.3 demonstrates the effect with a manipulation of Hickey's (1960,
1963) life table for sheep, as modified by Caughley (1967a). The rate of increase
implied by the $l_x m_x$ schedule is $r_s = 0·200$, and the stable age distribution is
that appropriate to this rate. Next to them is a second pair of l_x and S_x schedules
displaying the effect of an additional agent of mortality acting on all age classes
at an isolated rate of 0·1 per year. The new mortality rate q'_x for each age class
is therefore $q'_x = q_x + 0·1 - 0·1 q_x$ (Section 8.3) and from these a new survival
schedule l'_x has been calculated. Fecundity rates are unchanged. The rate
of increase implied by the $l'_x m_x$ schedule is $r' = 0·0945$. Yet

$$l_x e^{-rx} = l'_x e^{-r'x}$$

and the age distribution is therefore unchanged. This effect is of practical
importance in sustained yield harvesting.

Stable sex ratio

In a population with fixed schedules of survival and fecundity the ratio of males to females is given by

$$\frac{\male\male}{\female\female}=\frac{\male\sum l_x e^{-rx}}{\female\sum l_x e^{-rx}}$$

Even though the survival schedules of males may differ from that of females, the number of females and the number of males in the population increases at the same rate r, and the sex ratio therefore has no time trend. Further, the stability of the sex ratio is independent of the sign of r, of the birth rate, and of the sex ratio at birth.

Hanson (1963 : 58) considered that 'if differential survival continually tends to increase the number of adult males beyond approximately 150 per 100 adult females, a population of birds or mammals would eventually become extinct unless compensating production of young is very great.' Hanson apparently confused stability of numbers with stability of sex ratio. An example will illustrate the point. Consider a model population with zero rate of increase and a survival schedule common to males and females. Hence

$$\female\sum l_x m_x = 1$$

and

$$\frac{\male\male}{\female\female}=\frac{\male\sum l_x}{\female\sum l_x}=1$$

By some trick of habitat manipulation the survival of males is now greatly increased, survival of females and fecundity remaining unchanged. The sex ratio changes to a new value determined by the new male survivorship l'_x:

$$\frac{\male\male}{\female\female}=\frac{\male\sum l'_x}{\female\sum l_x}>1$$

and when the sex ratio reaches this level it stabilizes. Note that the relationship

$$\female\sum l_x m_x = 1$$

remains unchanged and hence the number of offspring added to the population each year is precisely the same as before the increase in survival of males. After the sex ratio has converged to its new level the rate of increase of the entire population will return to zero as before, the only differences being that the population is now larger and the sex ratio is disparate. In fact the sex ratio could climb to 10 or 100 without either causing instability or requiring a compensating change in fecundity of females.

Stable ratio of breeders and non-breeders

Lack (1966 : 293) discussed the model of a population of breeding birds limited by the number of nesting sites on an island. He considered that although the size of the breeding segment is limited, there would be 'a steady and continuing increase in the number of non-breeding individuals, up to a point where they would seriously deplete the food supply near the island, after which there would be density-dependent mortality from starvation.' Put in another way, Lack is suggesting that even though $r = 0$ for breeders, r for non-breeders will hold at a level above zero indefinitely, or until their survival is changed by an extrinsic agent of mortality.

At first glance Lack's model appears to violate two constancy laws of demography—that of a rate of increase common to all segments of the population and that of stable segmental proportions. The model can therefore serve as an alleged misdemeanour against which the power of the constancy laws may be tested. For this purpose the argument over the efficacy of density-dependent factors is totally irrelevant; we are interested only in whether Lack's conclusions on the dynamics of the system follow logically from the initial conditions and the processes that he has stipulated.

To simplify analysis we will assume that rates of mortality are constant with age. This assumption is not critical and conclusions do not depend upon it. Let

N = the fixed number of breeding birds
m = the number of offspring/year/breeding bird
n = the number of non-breeding birds
q = mortality rate of non-breeding birds. Those lost to the non-breeding segment because they leave it to replace losses in the breeding segment represent a component of this 'mortality'.

At the birth pulse the non-breeding segment comprises

$$n = Nm + Nm(1 - q) + Nm(1 - q)^2 + \ldots + Nm(1 - q)^\infty$$
$$= Nm(1 + (1 - q) + (1 - q)^2 + \ldots + (1 - q)^\infty)$$
$$= Nm\left(\frac{1 - (1 - q)^\infty}{q}\right)$$
$$= \frac{Nm}{q}$$

Hence the number of birds in the non-breeding segment at any birth pulse is determined exclusively by the values of N, m and q, each of which is a constant, not a variable. The number of non-breeders present at any birth pulse is therefore also a constant, and consequently the rate of increase of non-breeders is zero. If $N = 100$, $m = 3$ and $q = 0.6$·the stable number of non-breeders at the birth pulse is 500. Immediately before the next pulse it has declined to

$$\frac{Nm(1 - q)}{q} = 200$$

and it jumps again to 500 with the influx of new births. Lack's model conforms exactly to the constancy laws. His conclusions, not the laws, are at fault. Ecologists must train themselves to view apparent violations of these laws

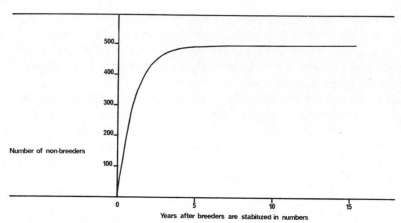

Figure 9.2.   The trend in numbers of non-breeding birds following stabilization of the number of breeding birds. See text.

with that degree of scepticism physicists reserve for blueprints of perpetual
motion machines.

Figure 9.2 indicates the speed with which constancy of segmental proportions
is achieved in the previous example. The size of the non-breeding segment
reaches 97 per cent of its equilibrium level only four years after the number
of breeders is stabilized. Full equilibrium is achieved five years later.

9.6 INTERPRETING AN AGE DISTRIBUTION

The age-distribution tautology

A tautology is a necessarily true statement that contains no information.
'If I had a mistress called Mary, her name would begin with M' is tautological.
It says nothing about my love life, neither revealing the name of my mistress
nor even whether, in fact, I have a mistress.

Tautologies also occur in mathematics. At first glance the equation

$$x = \log_e \frac{1}{2 \cdot 718^{-x}}$$

seems to be telling us something—it appears to attribute a numerical value
to x—but on solving we discover only that $x = x$. This is a mathematical
tautology.

Tautologies are prevalent in population analysis. The commonest is an
estimation of r, or a test of $r = 0$, from an age distribution. Sometimes the
calculation is run on the standing age distribution alone; more often the
distribution of ages at death and the age-specific fecundity schedule are treated
in combination with the age distribution. These exercises in circularity are

plays on the basic equation

$$\sum l_x e^{-rx} m_x = 1$$

It cannot be solved for r when l_x is calculated from the age distribution, S_x, on the assumption that $S_x = l_x$. Neither is a solving for r valid when the survival schedule is compiled from a distribution of ages at death, S'_x, on the assumption that $S'_x = d_x$. Both assumptions carry the correlate proposition that $r = 0$. A solving for r simply returns an estimate of this value. When the age distribution rather than the survival schedule is used in estimating r, the equation being operated on is not

$$\sum l_x e^{-rx} m_x = 1$$

but

$$\sum S_x e^{-rx} m_x = 1$$

The logical fallacy becomes obvious on substituting $l_x e^{-rx}$ for S_x to give

$$\sum (l_x e^{-rx}) e^{-rx} m_x = 1$$

The second r in this equation, on solving, necessarily returns $r = 0$ irrespective of the true value of r. Rate r calculated thereby will usually differ from the expected zero, but the deviation reveals nothing about rate of increase. It reflects sampling bias and sampling variation.

Tautologies need not be as subtle as this. Note these relationships:

$$\sum l_x e^{-rx} m_x = \sum S_x m_x = \sum d_x = l_0 = S_0 = 1.$$

They are true by definition. The value of r has no bearing on these equalities. A demonstration that $\sum d_x = l_0$, for instance, tells nothing about rate of increase although it has several times been presented as 'proof' that $r = 0$.

Variants of the age distribution tautology abound but they are difficult to recognize without close reading. An age distribution tautology cannot be detected unless the writer states exactly how he collected his data and how he analysed it. Solving for r might be valid when l_x is estimated by mark-recapture but invalid when estimated from either a shot sample of a population's age distribution or a picked-up sample of ages at death. Caughley and Birch (1971) list several examples of tautological analyses of this kind.

Table 9.4 demonstrates that two or more sets of superficially similar data can differ completely in their ability to estimate r. The data are part of a larger array (Lowe 1969) of age distributions of red deer on the island of Rhum, Scotland. Lowe constructed the age distributions by ageing animals found dead, thereby allowing an estimate of date of birth and period of survival for each specimen. As a very high proportion of total deaths was found, these data could be used to reconstruct the numbers alive in each age class in each of several years.

Four life tables could be constructed from the frequencies in Table 9.4.

Table 9.4. Age frequencies of female red deer on the island of Rhum. From Lowe (1969)

Year	Age in Years													
	0	1	2	3	4	5	6	7	8	9	10	11	12	13
1957	154	129	113	113	81	78	59	65	55	25	9	8	7	2
1958	131	129	129	110	107	69	75	51	59	43	12	5	6	3
1959			123											
1960				96										
1961					75									
1962						52								
1963							37							
1964								23						
1965									19					

1. From the 1957 standing age distribution on the assumption that $r = 0$,
2. from the 1958 standing age distribution on the assumption that $r = 0$,
3. from the yearly decline between 1957 and 1965 of the cohort born in 1957 (the diagonal of frequencies), and
4. from the difference between the number in a given year class in 1957 and the number in the next year class a year later.

Four separate life tables would result from these calculations but they cannot be combined with a fecundity table to provide four separate estimates of r_s. Only life tables 3 and 4 can be used in this way, the others being constructed on the assumption that $r = 0$. In combination with the fecundity schedule provided by Lowe (1969), life table 3 returns $r_s = 0.136$ and life table 4 returns $r_s = 0.082$. These are estimates of the rates of increase implied by the schedules of survival and fecundity. Because the age distribution was unstable they do not represent the actual rate at which the population increased at the times of measurement.

Fisheries biologists often interpret a log-linear decline of age frequencies as evidence that $r = 0$. Leaving aside the question of whether the stationary age distribution is typically of this form, log-linearity of frequencies is no basis for judging that $r = 0$. If q_x is constant for catchable ages the logs of the stable frequencies will regress linearly on age when rate of increase is zero. The regression will also be linear when the population is increasing or decreasing at any value of r.

Age distributions as indicators of r

The previous section outlined the reasons why r cannot be calculated from a standing age distribution together with a fecundity schedule. Whether the standing age distribution roughly indicates the sign of r is an entirely different question. In most instances it does not. In extreme cases, for instance at the

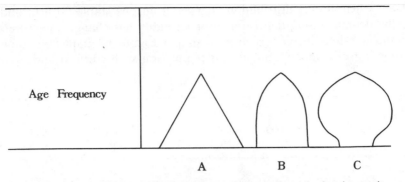

Age Frequency

A   B   C

Figure 9.3. Age distributions of urban populations of people. A = increasing population, B = stable population, and C = declining population (after Bodenheimer 1938).

breaking of a drought that almost entirely eliminated the yearling and two-year-old classes from a kangaroo population, a sudden change in age distribution can safely be interpreted as indicating a change in rate of increase. But then we would be reaching a conclusion that should be obvious from many other lines of evidence. More commonly, we seek the sign of r when r does not differ greatly from zero in either direction. Here, the age distribution seldom reveals whether r is positive, negative or zero, whether the distribution is divided into two or ten age classes.

That may seem a harsh judgement, particularly as the opposite conclusion is an axiom of wildlife research and management. The opposing view originated largely from misinterpretation of age distributions of urban man diagrammed by Bodenheimer (1938). He showed that age distributions of people differed according to whether the population was increasing, decreasing or stationary, and suggested that similar studies should be made for other species. Many subsequent writers have accepted these diagrams (Figure 9.3) as diagnostic of rate of increase for any species, mice no less than men. Populations of industrialized man are peculiar in that juvenile mortality is very low. The age distribution therefore differs from that of all other species. Diagram A could represent an increasing, stationary or declining population of most species of mammals or birds. The age distributions labelled B and C will be very rare in nature. They would represent an extreme rate of decline.

When a stable age distribution changes, one of three effects has occurred:

1. survival has changed over at least one age interval,
2. fecundity rate has changed at one or more ages, or
3. changes have occurred in both survival and fecundity.

But when an age distribution is not stable, a further change in age distribution implies no necessary change in either l_x or m_x. It may result simply from a convergence towards the stable age distribution appropriate to an $l_x m_x$ regime established some time before.

Even when an age distribution changes from one stable form to another, thereby demonstrating that survival or fecundity has changed, interpretation in terms of rate of increase is far from simple. Figure 9.4 graphs the stable age distribution of a model mammalian population with survival and fecundity schedules of:

Age:	0	1	2	3	4	5
l_x :	1·000	0·400	0·320	0·192	0·077	0·015
m_x :	0·0	0·6	1·2	1·4	1·3	0·5

A quick calculation will show that these schedules describe a stationary population with $r = 0$.

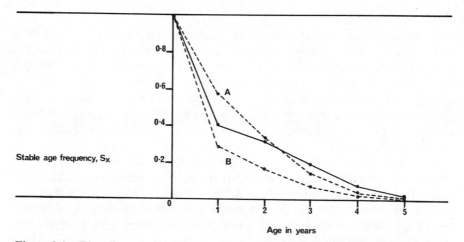

Figure 9.4. The effect on the stable age distribution of changing the rate of increase. A population with zero rate of increase (continuous line) changes to a new rate of $r = 0.33$ consequent on A: a doubling of first-year survival, B: a doubling of all fecundity rates.

The dashed lines give the stable age distribution that this population would eventually assume if the rate of increase rose suddenly to 0·33, consequent on

(A) a doubling of the survival rate over the first year of life, or
(B) a doubling of all fecundity rates.

Models A and B, although sharing a common rate of increase and a common origin, have different age distributions. The disparity reflects the contrasting strategies by which r was raised.

Disparity of the age distributions graphed in Figure 9.4 emphasizes the difficulty of interpreting an age distribution in terms of r. By no calculation or criterion can any of these three age distributions be linked unequivocally to a positive, negative or zero rate of increase. The difficulty is not diminished when the distributions are presented as age ratios (e.g. as juveniles to others, or as yearlings to adults). Unless evidence additional to the age distribution is available, a change in age distribution can seldom be interpreted in terms of r.

Although these remarks are fully justified, they are not offered in a spirit of pessimism. Bodenheimer (1938) showed that age distributions of man could be interpreted critically. Those of other species should be no less tractable. But the diagnostic features of age distributions of declining, stationary and increasing populations must be determined separately for each species or group of species. If the processes are known by which a population of a given species increases or decreases, interpretation of the age distribution is possible.

The processes responsible for change in rate of increase are partially understood for thar (Caughley 1970a). A positive rate of increase is generated by a rise in first-year survival and second-year fecundity. A decline is initiated by a drop in both first-year survival and adult fecundity. This pattern appears to be general for thar, and I suspect its generality extends to many other ungulate species. Such information allows cautious interpretation of the age distribution of thar because it sets a limit on the number of ways the age distribution can change. But without prior knowledge of this kind, interpretation of an age distribution is an exercise in clairvoyance.

9.7 GENERATION LENGTH

The length of a generation has little relevance to studies of populations of one species. Its usefulness is usually limited to comparisons between species, particularly when rates of change in gene frequencies are studied. A comparison between the rates at which mice and elephants evolve would make scant demographic or genetic sense (although it might make evolutionary sense) unless the change is expressed as a rate per generation. The length of a generation is thereby used as a scalar to compensate for the differing rates at which energy flows through populations of the two species.

The most common definition of mean generation length, T, is provided by the equation

$$R_0 = e^{rT}$$

where R_0, the 'net reproductive rate' or 'finite rate of increase per generation', is defined as

$$R_0 = \sum l_x m_x.$$

By substitution,

$$T = \frac{1}{r} \log_e \sum l_x m_x,$$

which seems rational enough until we solve for T when $r = 0$ and find that

$T = 0/0$. This value of T could be interpreted as one age interval. If the age interval were set at one year, T would equal one year. But if the same data were tabulated by intervals of two years, T would mysteriously increase to two years although the mortality and fecundity patterns remained unchanged. T is obviously not a useful measure of generation length.

Several writers (e.g. Andrewartha and Birch 1954, Cole 1960 and Laughlin 1965) suggested that a generation length can be defined as the mean interval between the birth of a mother and the birth of her offspring. Laughlin (1965) termed this 'cohort generation length, T_c' and I follow his usage. Reduction of the definition to algebra produces

$$T_c = \frac{\sum l_x m_x x}{\sum l_x m_x}.$$

It provides a logical measure of generation length for a cohort, and it is useful for this reason, but its utility flags when it is applied to a population consisting of several overlapped generations. In this situation the formula needs to be weighted by the proportions of animals in each age classes. An appropriate definition for this case is 'mean age of the mothers of all newborn females in a stable age distribution'. It is measured at a birth pulse as

$$\bar{T} = \frac{\sum l_x e^{-rx} m_x x}{\sum l_x e^{-rx} m_x}$$

$$= \sum S_x m_x x,$$

where, as before, S_x is the stable ratio of females aged x to newborn females. $\bar{T}$ is identical with T_c except that the temporal age distribution of a cohort (l_x) is replaced by the stable age distribution ($l_x e^{-rx}$). The concept of mean generation length $\bar{T}$ was proposed independently by Leslie (1966) and Goodman (1967). Since it is completely general, $\bar{T}$ should be used as the measure of mean generation length in all cases. Its analogue for a birth-flow population is

$$\bar{T} = L_x e^{-r(x + 1/2)} m_{x, x+1} (x + \tfrac{1}{2}).$$

Accepting $\bar{T}$ as the measure of mean generation length we can now rigorously define a finite rate of increase per generation (R) by

$$R = e^{r\bar{T}}$$

$$= \exp(r \sum S_x m_x x).$$

9.8 RATES OF BIRTH AND DEATH

The life table and fecundity table can be summarized by a pair of statistics, the birth rate and death rate.

Finite birth rate e^b will be defined (Caughley 1967a) as the rate at which a population would initially increase if deaths ceased. For a birth-pulse population with a stable age distribution

$$e^b = \frac{\sum l_x e^{-rx}}{(\sum l_x e^{-rx}) - 1}$$

$$= \frac{\sum S_x}{(\sum S_x) - 1} \cdot$$

It may come as some surprise that the equation defining birth rate carries no term for fecundity, but fecundity appears implicitly as a component of r. Yet even r need not be known because e^b can be solved directly from a stable age distribution.

No similar short-cut assists calculation of the finite death rate e^{-d}, which is defined by analogy with birth rate as the finite rate at which a population would initially decline if births ceased. When a population has a birth-pulse breeding system and a stable age distribution

$$e^{-d} = 1 - \frac{\sum d_x e^{-rx}}{\sum l_x e^{-rx}} \cdot$$

Although $\sum l_x e^{-rx}$ can be calculated from the stable age distribution, there is no way of getting $\sum d_x e^{-rx}$ from a distribution of ages at death unless r is known. Neither can $\sum d_x e^{-rx}$ be estimated from the stable age distribution because

$$d_x e^{-rx} \neq l_x e^{-rx} - l_{x+1} e^{-r(x+1)}$$

except in the special case of $r = 0$.

Finite rates of birth and death are related to finite rate of increase by $e^b \times e^{-d} = e^r$, the exponential rates being related by $b - d = r$.

Interpretation of b and d

A rise in the birth rate reveals only that more births are being produced per head, not necessarily that one or more values of m_x have increased or that r has increased. For instance, a decline in r consequent on increased juvenile mortality in one year will load proportionately more animals into the highly fecund age classes. The production of births per head will therefore rise even though the absolute number of births will probably drop. No causal feedback need be responsible for a rise in birth rate following a rise in mortality. The increased birth rate is an arithmetic consequence of what 'birth rate' measures: the interaction between age-specific fecundity rates and the age distribution. Consequently birth rate is as sensitive to changes in mortality as it is to changes in fecundity.

Neither is death rate purely a measure of mortality. A rise in fecundity rate will increase the proportion of juveniles in the population. These have a higher mortality rate than adults and the death rate therefore rises even when the l_x schedule remains constant.

For these reasons mortality and fecundity patterns should be examined in terms of l_x (or q_x) and m_x. Birth rates and death rates are useful in population analysis but provide ambiguous summaries of population processes. They express effects without revealing causes.

9.9 GROWTH OF POPULATIONS

This section has two aims: to show the pattern of population growth that results from various relationships between population parameters, and to provide the briefest of introductions to the techniques of modelling a population.

Models can be divided into two classes: general (or strategic) and special (or tactical). Strategic models dispense with the details of relationships within a population and between the population and its environment, concentrating instead on the dominant characteristics of the system. Their function is to reveal the general outcome of a process stripped to its essentials. The conclusions drawn from it are qualitative—whether the outcome is liable to be extinction, or stable equilibrium, or oscillations. The conclusions drawn from such a model are likely to hold true for most populations characterized by the modelled system of growth.

Tactical models, on the other hand, are designed to predict the detailed outcome, or the probability of this or that alternative outcome, for a highly specified set of initial conditions, parametal values and parametal relationships. A strategic model might indicate that there is a broad class of systems, to which the population of interest belongs, that almost invariably has a stable equilibrium point. That is useful information, but we might also need to know for our specific population whether the equilibrium is ever attained and the chance of the population extinguishing while hunting for that equilibrium point. These questions fall within the domain of tactical models which differ from strategic models in taking account of time lags, stochastic effects, fluctuating environments, spatial heterogeneity of habitat, dispersal, age distributions and seasonality. They simulate the specific reality of a given population growing in given conditions. Although they are useful for predicting what the population will do, they seldom provide general insights into ecological processes.

The two aims of this section are served best by strategic models. We will ask what is the deterministic outcome of a growth process in a constant environment when the perturbing effect of changing age distributions is ignored, where effect immediately follows cause, and where rates of reproduction and mortality are independent of age and season.

9.9.1 The first trophic level

Plants are usually limited by water and light, resources that are renewed at a rate independent of plant density. On a square metre of ground the resource comes in at a rate g and each unit of plant biomass must take up this resource at a rate b if it is to maintain itself and replace itself in the next generation. If we have a biomass V of vegetation on that square metre it can use a proportion of the resource equal to bV/g for replacement, leaving $1 - bV/g$ surplus which can be utilized for the population's growth. When the plants are sparse almost all available energy can be channelled into growth. As density increases so too does the proportion of the resource used for maintenance and replacement until finally all the resource is used in this way. The population then stops growing.

This system can be modelled approximately as

$$r_1 = r_{m1} \left(1 - \frac{bV}{g} \right)$$

The population's rate of exponential growth, r_1, is close to its intrinsic rate of increase, r_{m1}, when vegetation density is low, but is reduced progressively to $r_1 = 0$ as V increases. V finally stabilizes at a level of K where $K = g/b$. If K is substituted into the equation we arrive at

$$r = r_{m1} \left(1 - \frac{V}{K} \right)$$

which is the equation for logistic growth. Note that the constants for rate of renewal of the resource and rate of maintenance intake no longer appear;

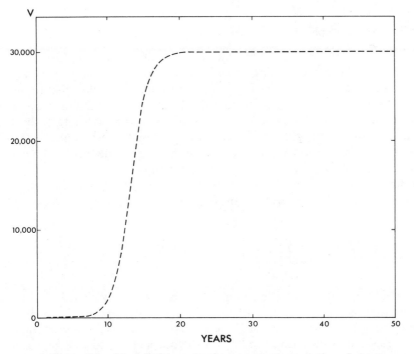

Figure 9.5. Modelled growth of an ungrazed population of plants.

they are now implicit components of the constant K. This substitution is possible only when rate of renewal of the resource is totally independent of the population's density, a property which with few exceptions is unique to the first trophic level. The growth of a population that influences the level of resources available to the next generation, as does a herbivore, cannot formally be modelled by a logistic equation. There are exceptions: an animal living on the flesh of stone fruit does not affect the amount of this resource available next year, and its population might therefore grow logistically. Secondly, the growth of populations that do influence the level of a resource available to the next generation but, having a low r_m, do not influence it profoundly, can be modelled by the logistic equation as a pragmatic approximation. The population's trajectory will differ from the logistic trajectory but the fit will often be close enough for practical purposes.

Figure 9.5 shows the trajectory of a population growing under logistic rules, parametered by $r_{m1} = 0.8$ and $K = 30,000$. The biomass of grass on your back lawn would have a similar trajectory if seed were planted in bare soil and the grass allowed to grow free of the influence of a lawn mower. Of course it would not follow a logistic curve exactly because the rate at which resources are renewed is fluctuating rather than constant, and because temperature fluctuates and age distribution changes, but it would be close enough. The population in Figure 9.5 is of a slower growing species, a shrub that requires about 20 years to reach equilibrium density.

9.9.2 The second trophic level

A population in the second trophic level is subjected to influences very different from those determining growth in the first level. Its growth rate is influenced by the standing crop of its limiting resource, and when that resource is food the population itself influences the level of that standing crop. The system is circular: the standing crop of food is determined largely by the animals and the standing crop of animals largely by the availability of food. Hence we need two equations, one for rate of increase of the vegetation and a second for rate of increase of the herbivore population.

The first equation will be that given previously for rate of increase of vegetation, modified by a term expressing the degree to which this rate is reduced by herbivores (H):

$$r_1 = r_{m1}\left(1 - \frac{V}{K}\right) - c_1 H(1 - e^{-d_1 V})/V.$$

The constant c_1 is the rate of consumption of vegetation by a single herbivore when its food is unrationed (i.e. when V is high and therefore $1 - e^{-d_1 V}$ approaches unity). At lower levels of V the animal cannot take in a satiating diet and c_1 is reduced by the term $1 - e^{-d_1 V}$ which approaches zero as V approaches zero. The constant d_1 determines the rate of fall from 1 to 0. Thus the term

$c_1(1 - e^{-d_1 V})$ models the herbivore's 'functional response', the response of its rate of intake to the availability of food.

Next we require an equation for the herbivore population's rate of growth:

$$r_2 = -a_2 + c_2(1 - e^{-d_2 V})$$

where a_2 is the rate of decrease per head in the absence of food and c_2 is the rate at which this decline is ameliorated when food is abundant. The population's intrinsic rate of increase is therefore $r_{m2} = c_2 - a_2$. With declining levels of food c_2 is reduced progressively by $(1 - e^{-d_2 V})$ which takes values between 0 and 1 according to the level of V. The term $c_2(1 - e^{-d_2 V})$ models the herbivore's 'numerical response', its reaction to density of food in terms of survival and reproduction. The vegetation–herbivore system is therefore modelled by the linked equations.

$$r_1 = r_{m1}\left(1 - \frac{V}{K}\right) - c_1(1 - e^{-d_1 V})/V$$

$$r_2 = -a_2 + c_2(1 - e^{-d_2 V})$$

Figure 9.6 shows these equations in action. The graph starts with the plant population of Figure 9.5 at equilibrium, and into this garden of Eden we inject twenty herbivores with these characteristics: $c_1 = 2 \cdot 5$, $d_1 = 0 \cdot 0001$, $a_2 = 1 \cdot 2$

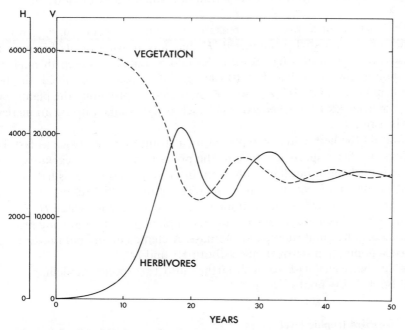

Figure 9.6. Modelled growth of a population of herbivores and the trend of biomass for the plant population on which it feeds.

and $c_2 = 1\cdot7$ and $d_2 = 0\cdot00008$. The population has been grown in a programmable desk calculator by estimating growth rates twenty times per year and adding on the appropriate increment each time. For this exercise the constants representing yearly rates (r_{m1}, c_1, a_2 and c_2) are divided by the number of iterations per year. Most wildlife managers will recognize the trajectories of Figure 9.6 as a blue-print for an ungulate eruption and we can therefore be cautiously confident that the model is duplicating the essential features of a vegetation–herbivore system. It reproduces the pattern of growth we see in the field.

There is no necessity to 'grow' the population to discover the general outcome. It can be estimated analytically. R. M. May and E. E. Robinson have analysed the stability of this model and have kindly allowed me to use their results. In brief, if an accommodation is possible between the vegetation and the hervibores the biomass of these two components of the system will move to an equilibrium at

$$V^* = \frac{1}{d_2} \log_e \left(\frac{c_2}{c_2 - a_2} \right)$$

$$H^* = \frac{V^* r_{m1} (1 - V^*/K)}{c_1 (1 - e^{-d_1 V^*})}$$

Those two equation allow general insights into the workings of a vegetation–herbivore system which are neither trivial nor immediately obvious.

1. Neither intrinsic rate of increase of the edible vegetation, r_{m1}, nor its equilibrium biomass when ungrazed, K, has any bearing on its biomass V^* at equilibrium with grazing pressure. Should a mutation sweep through the plant population such that its intrinsic rate of increase is doubled, the new equilibrium is V^* and $2H^*$. Likewise K increased by providing the plants with more water leaves V^* unchanged. The slack is again taken up by an increase of herbivores.

2. Should the herbivores be replaced by a strain that eats twice as fast, but which is in other respects identical to the previous strain, the vegetative equilibrium is unchanged but the equilibrium biomass of herbivores is halved.

3. A change in the intrinsic rate of increase of the herbivores ($r_{m2} = c_2 - a_2$) affects both V^* and H^*, a rise resulting in reduced V^* and increased H^*.

4. In general, a change in any parameter of plant growth or grazing pressure has no effect on equilibrium plant biomass. A change in any parameter of the herbivore population's growth affects both V^* and H^*.

Should the reader wish to push farther into two-species modelling, May's (1973) book is essential reading.

9.9.3 The third trophic level

By now the strategy of modelling a population's growth will be apparent.

We need one equation for growth of the population and at least one for each trophic level below it. A very simple system can be modelled as

$$\text{Vegetation}: \quad r_1 = r_{m1}\left(1 - \frac{V}{K}\right) - c_1 H(1 - e^{-d_1 V})/V$$

$$\text{Herbivores}: \quad r_2 = -a_2 + c_2(1 - e^{-d_2 V}) - fP(1 - e^{-d_3 H})/H$$

$$\text{Predators}: \quad r_3 = -a_3 + c_3(1 - e^{-d_4 H})$$

P in the second equation is the standing crop of predators and the term in which it occurs is the predators' functional response, equivalent logically to the functional response of herbivores in the first equation. The system modelled by these equations may either have a stable equilibrium or it may oscillate, according to the values of the constants. This is precisely what we see in the field. Some vegetation–herbivore–predator systems are relatively stable while others, particularly at high latitudes, are oscillatory.

Figure 9.7 shows what happens when twenty predators are introduced into the system diagrammed in Figure 9.6. The additional constants are assigned values of $f = 2.0$, $d_3 = 0.005$, $a_3 = 1.0$, $c_3 = 1.4$, and $d_4 = 0.005$. Vegetation biomass rises, and herbivore biomass falls in response to an increasing population of predators. The outcome is a new equilibrium. My reading of the Isle Royale story (Mech 1966, Allen 1974), where moose established and erupted and then wolves arrived and preyed upon them, is that Figures 9.6 and 9.7 simulate in kind the history of events. By this interpretation the moose were

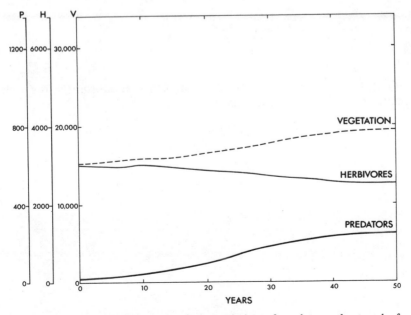

Figure 9.7. Modelled growth of a population of predators, the trend of herbivores on which it feeds, and the trend of plants sustaining the herbivores.

approaching an equilibrium with the availability of food when the wolves arrived and diverted them to a lower equilibrium.

Before we leave predators we will look at a peculiarly unsubtle one—modern man. Suppose he preys upon a population of herbivores, his effort being regulated by the issue of a constant number of hunting permits each year but with no further restrictions being imposed on a permit holder. For further simplification we assume that each permit holder is an efficient professional hunter backed by a staff that will process his harvest. In these circumstances the system grows as

$$\text{Vegetation}: \; r_1 = r_{m1}\left(1 - \frac{V}{K}\right) - c_1 H(1 - e^{-d_1 V})/V$$

$$\text{Herbivores}: \; r_2 = -a_2 + c_2(1 - e^{-d_2 V}) - fP$$

$$\text{Predators}: \; r_3 = 0.$$

The functional response of predators per head of herbivores is reduced from the $fP(1 - e^{-d_3 H})/H$ of the previous set of equations to fP. Thus the number of herbivores the predator takes is now only a function of his efficiency and the number of herbivores available. He will take the same proportion of the population at high or low density, there being no satiating diet to which his harvest converges. The equilibrium of this system is

$$V^* = \frac{1}{d_2} \log_e\left(\frac{c_2}{c_2 - a_2 - fP}\right)$$

$$H^* = \frac{V^* r_{m1}(1 - V^*/K)}{c_1(1 - e^{-d_1 V^*})}$$

$$P^* = P.$$

We will return to the practical implications of this equilibrium in the section of Chapter 11 dealing with sustained yield.

Chapter 10

Mark–recapture

Mark–recapture techniques have a long history, a plethora of designs and analyses, and a literature of daunting size. Fortunately, a very complete summary has been published by Seber (1973) which, together with Cormack's (1968) able review, provides an introduction to a broad range of designs and analyses, some of which may be needed for unusual problems. A less detailed but very useful review is provided by Hanson (1967).

This chapter covers only those designs of immediate interest to vertebrate ecologists.

All mark–recapture experiments have the same basic format. A sample of animals is captured, marked and released again, the properties of this identifiable sample then being used to estimate the properties of the population as a whole. The aim of the exercise would be to investigate one or more of these properties:

1. movement
2. growth rate
3. age-specific fecundity rates
4. age-specific mortality rates
5. size of the population
6. rate of birth and immigration combined
7. rate of death and emigration combined
8. rate of harvesting
9. rate of increase.

The use made of the marked sample depends on the purpose of the experiment. An experimental design appropriate to one aim may be entirely inadequate for another. Experimental design is critical. Too often, animals are marked and recaptured before the biologist has decided what information he seeks; most biometricians can relate atrocity stories of this kind. Before launching into an experiment the biologist must decide on the parameters he needs to estimate, the experimental design appropriate to this aim, and the level of precision he requires. It may be that mark–recapture is not needed, that the required information can be extracted by simpler methods. If another technique will suffice it should be used. Mark–recapture is both time consuming and expensive, and the results are often inaccurate because mark–recapture models are seldom more than a vague approximation to reality.

10.1 CATCHABILITY

Mark–recapture experiments fall into two classes, those appropriate to aims 1–4 of the previous classification and those appropriate to 5–9. Analyses in the first class require only that marked individuals can be identified. Only the marked sample is studied. The behaviour and number of animals in the unmarked segment of the population is irrelevant to estimating parameters. We are not interested in whether marked individuals are more easily captured than unmarked individuals, although we may need to assume that marked individuals themselves are equally catchable. The second class of experiments comprises those in which the equal catchability of all individuals, marked and unmarked, is crucial to the accuracy of the results. No mark–recapture experiment of this kind should be initiated until the truth of the assumption has been tested in a pilot experiment, or unless the experimental design has such a test built into it.

Many writers, in presenting their mark–recapture results, confide that the assumptions underlying the analysis were probably violated to some (unmeasured) extent, but that they have done the best they can; and having made this ritualistic obsequiance towards statistical propriety they then proceed to interpret the results as if no possibility of error existed. Double-think of this kind is dangerous. Mark–recapture analyses are not particularly robust and small deviations from their implicit assumptions can produce large errors in the results. The greatest source of error is unequal catchability. Eberhardt (1969b), adding to a scheme proposed by Cormack (1966), classified its causes under three heads:

1. a property inherent in the individual (expressed in its behaviour in the immediate vicinity of a capturing device),
2. the result of learning (animals may become capture-prone or capture-shy), and
3. a property depending on relative opportunity of capture (an animal cannot be trapped if no trap is placed within its home range).

The first effect is difficult to detect from recapture records since in the extreme case some individuals will have a catchability of zero. No clue to the presence of these animals is available from the results of the experiment. The second effect is easily detected if animals are recaptured on more than one occasion (Section 10.1.1). Its antidote is to change the method of capture to one in which the chances of capture are less dependent on a decision made by the animal, to use a model that can cope with capture-proneness and capture-shyness

(Section 10.6), or to reduce the number of recapturing occasions so that no animal has time to learn much about how to get caught or how to avoid being caught. The third cause of unequal catchability—a property depending on relative opportunity of capture—is probably the most important and certainly the one receiving the least attention. The reliability of analyses acting on the estimated ratio of marked to unmarked individuals depends not so much on a random mixing of the two classes, a virtual impossibility anyway, as it depends on animals being marked at stations located at random throughout the range of the population or being recaptured at randomly located stations. Mark–recapture experiments on small mammals are often flawed because animals are recaptured at the same stations at which they were marked. Unless the trapping grid is very tight, so tight that every home range is occupied by at least one trap, failure to re-randomize trapping stations between marking and recapturing occasions results inevitably in an excessively high rate of recapture of marked animals, and a consequent underestimate of the size of the population. This result can be avoided by placing traps at random on only about 25 per cent of grid intercepts at the marking occasion. Before the recapture occasion the traps are moved to a second set of grid intercepts determined by a table of random numbers. By chance, of course, some of the grid intercepts will be occupied by a trap on both occasions. When the experimental design calls for more than one recapture sampling the capture stations must be randomized anew for each occasion.

This procedure fully randomizes opportunity of capture only when all individuals have the same radius of movement. If, for instance, males are more mobile than females they are more likely to encounter traps. Such heterogeneity of behaviour requires that the number of males is estimated separately from the number of females.

Unequal catchability, whatever the cause, is more the rule than the exception. It has been established for populations of mammals (Getz 1961, Kotte 1965, Tanton 1965 and 1969, Marten 1970), birds (Orians and Leslie 1958, Taylor 1966) and frogs (Turner 1960a). One might expect capture-shyness to be uncommon in fish, but Beukema (1970) showed that catchability of carp one year after being hooked and lost was three times lower than carp that had not previously been hooked.

10.1.1 Tests for equal catchability

Roff (1973) examined the commonly used tests for equal catchability and showed that they could not distinguish between a population in which probability of capture was truly random and one comprising two or more classes of animals differing in mean catchability. The tests do not distinguish between a compound Poisson distribution and a simple Poisson, and in the first case will indicate that animals are equally catchable when they are not. A significant result from any of these tests can confidently be interpreted as indicating unequal catchability; a non-significant result is ambiguous. We therefore apply these

Table 10.1. Data for Leslie's test of equal catchability (from Orians and Leslie 1958)

Year	Recaptures (n)	Times recaptured (i)	No. of animals (f)
1947	7	0	15
1948	7	1	7
1949	6	2	7
1950	4	3	2
1951	7	4	1
		5	0
	$\sum n = 31$		$\sum f = 32$
	$\sum n^2 = 199$	$\sum fi = 31$	$\sum fi^2 = 69$

tests to detect certain kinds of unequal catchability, rejecting those data that yield a significant result, and apply mark–recapture analysis only to those data for which unequal catchability, although perhaps occurring, cannot be demonstrated.

Suppose animals were marked on a single occasion and recapture samples were taken on six subsequent occasions. All the marked animals recaptured on the last of these have obviously been alive and available for recapture on each of the previous five recapturing occasions. The number of these animals that were recaptured once, twice and so on up to five times provides the data necessary to test for randomness of recapture (i.e. equal catchability).

Table 10.1 gives a set of data of this kind. It is a distribution of recapture frequencies for 32 shearwaters marked in 1946 and known to be alive in 1952 (Orians and Leslie 1958). Recapture samples were taken on five occasions over this period and therefore each individual could have been recaptured a maximum of five times. Leslie pointed out that if catchability is constant the recapture frequencies will form a binomial distribution. The hypothesis can therefore be tested by comparing the observed variance with the expected binomial variance. The test may be run as a χ^2 with $(\sum f) - 1$ degrees of freedom:

$$\chi^2 = \frac{\sum fi^2 - \dfrac{(\sum fi)^2}{\sum f}}{\dfrac{\sum fi}{\sum f} - \dfrac{\sum n^2}{(\sum f)^2}}$$

$$= \frac{69 - \dfrac{31^2}{32}}{\dfrac{31}{32} - \dfrac{199}{32^2}}$$

which gives $\chi^2 = 50.32$ with d.f. $= 31$ to yield $P < 0.02$. The hypothesis that the distribution of recaptures is binomial, and therefore that catchability is constant, is clearly not supported by this result.

Most χ^2 tables stop at 30 degrees of freedom (d.f.). If a test has a greater number the significance of the resultant χ^2 value can be tested by converting it to the standardized normal variate $\sqrt{(2\chi^2)} - \sqrt{(2 \text{ d.f.} - 1)}$ which, in the example, gives $\sqrt{(100.6)} - \sqrt{(61)} = 2.22$. When this value exceeds 1.96, P is less than 0.05 and the assumption of equal catchability is called into question.

Leslie's test has a grave practical disadvantage in that it rejects information from the first and last capturing occasions together with all the recapture records of animals not caught on the last occasion. The culling is necessary to eliminate from the analysis all but those animals known to be alive throughout the course of the experiment. Consequently the test may utilize less than a tenth of the total records. Unless the study is on a large scale these data may be too few for a meaningful test of equal catchability. However, if the period between first and last capturing occasions is short enough to ensure that natural mortality is low or zero, all capture records can be used in a test. The records are summarized as in Table 10.2 which gives a frequency distribution of animals captured $i = 1, 2, 3 \ldots$ times. Since all catching occasions are included, each animal has been caught at least once. The table does not therefore contain a zero frequency as does the data used in Leslie's test. If catchability is constant the distribution of frequencies will form a zero-truncated binomial distribution, which approximates a zero-truncated Poisson distribution when the number of sampling occasions is large relative to the mean number of times an animal is captured. Darroch (1958) warns that the expected distribu-

Table 10.2. Observed frequency of capture of agamid lizards and the zero-truncated Poisson frequencies to be expected if catchability is constant (J. A. Badham, unpublished)

Number of times captured	Number of individuals	Expected frequencies	$\dfrac{[f - E(f)]^2}{E(f)}$
i	f	$E(f)$	
1	23	15.797	3.284
2	7	12.303	2.286
3	3⎫	6.388⎫	
4	2⎪	2.488⎪	
5	1⎪	0.775⎪	
6	0 ⎬8	0.201 ⎬9.908	0.367
7	1⎪	0.045⎪	
8	0⎪	0.009⎪	
9	1⎭	0.002⎭	
10	0	0.000	
	$\sum f = 38$		$\chi^2 = 5.937$
			$d.f. = 1$
			$P = 0.02$

tion is never precisely a truncated Poisson but for present purposes it will serve as an adequate approximation. A truncated Poisson distribution is fitted to the observed frequencies and the fit is tested by χ^2. Expected frequencies are calculated from the mean, $\bar{X}$, of the complete Poisson distribution which is related to the mean of a zero-truncated Poisson by

$$\bar{x} = \frac{\bar{X}}{1 - e^{-\bar{X}}}$$

The mean of the observed frequency distribution of capture is equated with x and from this $\bar{X}$ is estimated. Table 10.3 is a table of conversions. If greater accuracy is required, $\bar{X}$ can be calculated from the above equation by trial and error using the nearest entry in Table 10.3 as a starting value and taking the exponential of $\bar{X}$ from Appendix 1. When $\bar{x} > 2$ the Lagrange series provides a direct solution (Irwin 1959) as

$$\bar{X} = \bar{x} - Z - Z^2 - 1 \cdot 5 Z^3 - 2 \cdot 6 Z^4 - 5 \cdot 21 Z^5$$

in which $Z = \bar{x} e^{-\bar{x}}$.

Table 10.2 comprises a distribution of the frequency with which 38 agamid lizards (*Amphibolurus barbatus*) were captured on 21 occasions over a period of 37 days. The distribution has a mean of $\bar{x} = \sum fi / \sum f = 1 \cdot 974$. From Table 10.3 it is obvious that $\bar{X}$ will be a little more than $1 \cdot 5$; it is actually $1 \cdot 558$.

The expected frequencies $E(f)$ for each i are now calculated as

$$E(f_1) = \frac{(\sum f) e^{-\bar{X}}}{1 - e^{-\bar{X}}} \cdot \frac{\bar{X}^i}{i!}$$

the first term being a constant for all values of i. The expected frequency at $i = 1$ is therefore

$$E(f_1) = \frac{38 \times 0 \cdot 2107}{0 \cdot 7893} \times 1 \cdot 558 = 15 \cdot 8$$

and each subsequent expected frequency is calculated by multiplying the previous frequency by $\bar{X}/i$. Hence

$$E(f_2) = 15 \cdot 8 \times \frac{1 \cdot 558}{2} = 12 \cdot 3$$

Table 10.3. Relationship between the mean $\bar{X}$ (in body of table) of a Poisson distribution and the equivalent mean $\bar{x}$ of the zero-truncated Poisson distribution

$\bar{x}$	0	1	2	3	4	5	6	7	8	9
1.	—	0·194	0·376	0·550	0·715	0·874	1·027	1·175	1·318	1·458
2.	1·594	1·726	1·856	1·983	2·109	2·232	2·353	2·472	2·590	2·706
3.	2·821	2·935	3·048	3·160	3·271	3·381	3·490	3·599	3·707	3·814
4.	3·921	4·027	4·133	4·238	4·343	4·447	4·551	4·655	4·759	4·862
5.	4·965	5·068	5·170	5·273	5·375	5·477	5·579	5·680	5·782	5·883

and

$$E(f_3) = 12{\cdot}3 \times \frac{1{\cdot}558}{3} = 6{\cdot}4$$

and so on.

The calculated expected frequencies (Table 10.2) are now tested for congruency with the observed frequencies of capture. The last eight frequencies are pooled to ensure that all expecteds exceed a value of 5, thereby reducing the distribution of observed and expected frequencies to three classes each. The χ^2 of 5·937 (Table 10.2) has only one d.f., two less than the number of classes in the test. Its associated probability of $P = 0{\cdot}02$ indicates that the observed frequencies are almost certainly not from a Poisson distribution. We conclude therefore that either animals are not equally catchable or that a significant number of them died during the experiment. Since the experiment was designed specifically to exclude the second alternative we accept the first as true.

Additional tests of equal catchability are given by Leslie (1952) and by Cormack (1966).

10.1.2 Testing for loss of marks

No statistic calculated by mark–recapture is accurate unless the marked animals stay marked. Unfortunately, very few of the marking systems in use are permanent. The worst offender is probably the bird band. It sometimes wears out and drops off (Blake 1951, Hickey 1952, Orians and Leslie 1958, Rowley 1966, Fordham 1967), is occasionally removed by the bird (Lovell 1948) and often it is lost at the fledgling stage because of poor fitting.

Farner (1949) suggested that the proportion of double-banded birds that lose one band could be used as an index of band loss. His index can be elaborated into a direct estimate of the rate at which bands are lost. A group of fledglings is banded on both legs with distinctive bands that do not differ structurally from bands used in routine single-banding. A sample of birds observed t units of time later will comprise four categories:

a. B_2 birds that retain both bands,
b. B_1 birds that retain only one of the original bands,
c. B_0 birds that retain neither band, and
d. birds that are not members of the double-banded cohort.

Only birds in categories a and b can be identified as such.

The probability that a single-banded bird will lose that band over t time units is designated p, and the probability that it will retain the band is therefore $1 - p$. If the loss of one band is independent of the loss of the other, a double-banded bird faces the probability p^2 of losing both bands, $2p(1 - p)$ of losing one or other of the two bands, and $(1 - p)^2$ of losing neither band over t time units. Hence

$$p^2 : 2p(1-p) : (1-p)^2 = B_0 : B_1 : B_2.$$

Suppose an observed sample contained $B_1 = 32$ single-banded birds, $B_2 = 64$ double-banded birds, and a remainder that carry no bands and therefore could not be classified according to whether or not they were originally double-banded. The number of these that have lost both bands is estimated by

$$B_0 = \frac{B_1^2}{4B_2} = \frac{32^2}{4 \times 64} = 4 \text{ birds.}$$

The probability that a bird originally banded with only one band would lose that band over the same period is

$$p = \frac{B_1}{2B_2 + B_1} = \frac{32}{128 + 32} = 0 \cdot 2.$$

The estimate of p can be used to correct records of recapture of birds banded originally with one band. If a recapture sample contains m banded birds, the number of birds in this sample that were originally banded, some of which have lost their bands, is

$$m' = \frac{m}{1 - p}.$$

When mortality rate is estimated by mark–recapture p must be estimate for several values of t, the period between marking and recapturing, to determine the trend of p with time.

This correction is an approximation, not an exact solution. The inexactness stems mainly from the underlying assumption that the loss of one band is independent of the loss of another. It is unlikely to be true in all cases.

10.2 INTERPRETATION OF POPULATION SIZE ESTIMATED BY MARK–RECAPTURE

When a population has distinct and known geographic bounds we ask only how an experiment should be designed to estimate the population's size. Often however, the population has no distinct boundary and we are limited to describing that part of it living on a study area of arbitrary size. If a mark–recapture estimate is used the question posed by the first instance—how do I measure this population?—becomes in the second: what is the population that I am measuring? Suppose that the population is sampled by a single line of closely spaced traps in the middle of the study area and that the traps are not moved between the first and second capturing occasions. The resultant estimate is not of the number of animals in the study area but of the number of animals whose home ranges overlap the trap line. If the line has a length of L metres and the mean diameter of a home range is D metres, the mark–recapture experiment estimates the population on a strip totalling about $D(D + L)$ square metres. If, however, the distance between neighbouring traps is considerably greater than the mean diameter of a home range, we have estimated population size on a set of discontiguous areas totalling about $n\pi(\frac{1}{2}D)^2$ square metres, where n is the number of traps. Now suppose that the design calls for marking at a number of traps located at random throughout the study area and that animals are recaptured after a second randomization of trap locations. The population is closed in the sense that immigration equals emigration. The population determined by a Petersen estimate (one marking, one recapturing) is that occupying the study area and a strip around its periphery whose width is about half the mean diameter of a home range. Further, the estimate will be too high even for this total area because although emigration has no effect on the estimate immigration increases it.

These problems of interpretation are encountered whenever a mark–recapture experiment is applied to a population without natural boundaries. The designing of a mark–recapture experiment to estimate the number of ducks in North America poses no technical problems but it is virtually impossible to design one leading to an accurate census of ducks in the United States.

10.3 PETERSEN ESTIMATE

The simplest mark–recapture estimate of numbers calls for marking on one occasion and recording the proportion of marked animals in a sample captured on a second occasion. The estimate is underpinned by the primary axiom of sampling theory: that the proportion of entities in the population that have a certain characteristic can be estimated from the proportion of these in a sample of the population. Within the limits of sampling variation

$$\frac{M}{N} = \frac{m}{n}$$

where M animals are marked in a population of size N (N being unknown) and m marked animals are recaptured in a subsequent sampling of n animals. It takes little manipulation to convert this equation to a form estimating population size:

$$N = M \left/ \frac{m}{n} \right. \quad \text{or} \quad N = \frac{Mn}{m}$$

i.e. the population size *at time of marking* is estimated by the number marked divided by the proportion of marked animals in a sample taken at a later date. The calculation yields a Petersen estimate of population size.

Mortality does not affect the estimate providing the rate of mortality of marked and unmarked animals is the same. If 70 per cent of animals died between marking and recapturing the equation could be written

$$N = M \left/ \frac{m(1 - 0\cdot7)}{n(1 - 0\cdot7)} \right.$$

with the survival rates $(1 - 0\cdot7)$ cancelling.

The Petersen equation is the basis of most mark–recapture estimates of numbers. We associate it so closely with marking and recapture that it comes as a surprise to learn that it was used first to estimate a population whose members were neither marked nor captured. Laplace (1786), quoted by Cormack (1968), estimated the population of France by dividing the registered number of births, M, recorded for the whole country over a year by that fraction of the population of certain parishes of known size n whose names appeared as newborn in the birth registers of that year. Thus registered newborn were treated as marked individuals and the proportion of these 'recaptured' in a sample of known size was used to estimate the total population. Laplace did not need to assume that the birth records were accurate (they were not), only that the proportion of births reported from his sampled parishes was the same as the proportion reported from the country as a whole.

The notion of using marked individuals to estimate numbers was first advanced by Petersen (1896) but was not put into practice until Dahl (1919) estimated the size of a fish population in this way. The estimate is named for Petersen in acknowledgement of his advocation of the method. Lincoln (1930) used it to estimate duck numbers, and wildlife managers, ornithologists and mammalogists often call it a Lincoln index. The name is unfortunate both because the method yields an estimate, not an index, and because Lincoln was a latecomer to the field. In the interests of standardization, if for no other reason, 'Petersen estimate' should be used to identify the method.

The Petersen estimate yields an accurate result only when these conditions are met:

1. the probability of capturing an individual is the same for all individuals in the population,
2. no animal is born or immigrates to the study area between marking and recapturing,
3. marked and unmarked individuals die or leave the area at the same rate, and
4. no marks are lost.

The intuitively reasonable relationship

$$N = \frac{Mn}{m}$$

actually produces a biased estimate that, over the long run, results in over-estimation of N (Chapman 1951; Bailey 1951, 1952). Bailey suggested that when the number of marked individuals to be recaptured is not decided prior to recapturing (direct sampling) a more satisfactory estimate is

$$N = \frac{M(n+1)}{m+1}$$

which has a formal standard error of approximately

$$\text{S.E.} = \sqrt{\left\{ \frac{M^2(n+1)(n-m)}{(m+1)^2(m+2)} \right\}}.$$

However if the number of recaptures is decided on before recapturing commences and recapturing continues until this number of marked individuals is caught (inverse sampling) then a completely unbiased estimate of population size is provided by

$$N = \frac{n(M+1)}{m-1}$$

with a standard error or

$$\text{S.E.} = \sqrt{\left\{ \frac{(M-m+1)(N+1)(N-M)}{m(m+2)} \right\}}$$

The disadvantage of inverse sampling is the need for a prior decision on the number to be recaptured. The biologist may set himself a target of m recaptures that he cannot reach in practice, particularly if he has no idea of the population's size. The advantages of inverse sampling are so great, however, that pains should be taken to acquire the information necessary to make it work.

The required precision of an estimate of population size should be decided upon before the recapture sample is taken. The number of marked animals that must be recaptured to afford this precision can be calculated from the number of marked animals known to be in the population together with a guess at the size of the population. Suppose a biologist suspected that a population contained about 200 members and that he knew 40 of these were marked. He required an estimate of numbers whose standard error is about 10% of the population's size—about 20. The number of marked individuals that he must recapture to provide a standard error of about this size can be approximated by

$$m = \frac{\sqrt{\left\{ \frac{(N+1)(N-M)}{\text{S.E.}^2} + 2 \right\}^2 + \left\{ \frac{4(N+1)(N-M)(M+1)}{\text{S.E.}^2} \right\}} - \left\{ \frac{(N+1)(N-M)}{\text{S.E.}^2} + 2 \right\}}{2}$$

$$= \frac{\sqrt{\left(\frac{201 \times 160}{400} + 2 \right)^2 + \left(\frac{4 \times 201 \times 160 \times 41}{400} \right)} - \left(\frac{201 \times 150}{400} + 2 \right)}{2}$$

$= 29 \cdot 45$, say 30 marked individuals.

Hence the prior judgement on the required precision of the estimate dictates that recapturing must continue until 30 marked individuals are recaptured.

We will suppose that the biologist now goes ahead and recaptures 30 marked individuals, in the process also capturing 155 unmarked individuals. Hence

$$M = 40$$
$$m = 30$$
$$n = 30 + 155 = 185.$$

The population size is estimated by Bailey's equation as

$$N = \frac{n(M + 1)}{m - 1}$$

$$= 185 \times 41/29 = 262 \text{ animals.}$$

This estimate has a standard error of

$$\text{S.E.} = \sqrt{\left\{ \frac{(M - m + 1)(N + 1)(N - M)}{m(m + 2)} \right\}}$$

$$= \sqrt{\left\{ \frac{(40 - 30 + 1) \times 263 \times 222}{30 \times 32} \right\}} = 26,$$

and so $N \pm \text{S.E.} = 262 \pm 26$ or 262 ± 10 per cent.

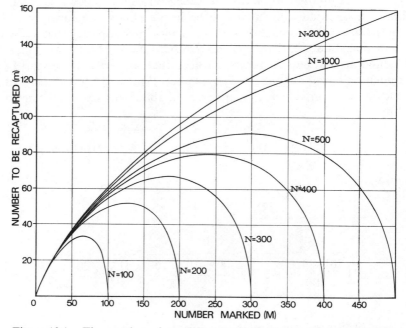

Figure 10.1. The number of marked animals that must be recaptured to provide a standard error that is 10 per cent of the Petersen estimate of the population. This number is a function of the number marked (M) and the population size (N).

Figure 10.1 shows for various combinations of population size and number marked the number of marked individuals that must be recaptured to give a standard error of around 10 per cent of the population size. The curves indicate that unless more than a third of the population is marked (an unlikely possibility) the accuracy of the guessed total is not critical. A mistake in the order of double or half will raise or lower the standard error by only a few per cent. The following combinations of M and m are an adequate general guide for a population containing between 100 and 1000 members that must be estimated with a standard error of 10 per cent or lower:

$$
\begin{array}{cccc}
M & : & 100 & 200 & 250 \\
m & : & 55 & 80 & 90
\end{array}
$$

The Petersen estimate is biased upward by birth and immigration. For this reason its use is best restricted to populations sampled between birth pulses. Bias accruing from immigration can be reduced by restricting the period between marking and recapturing. If individuals are born during the experiment an unbiased estimate of N demands data from at least three capturing occasions. Sections 10.5 and 10.6 outline designs and analyses appropriate to this case.

10.4 SCHUMACHER'S METHOD

The Petersen estimate calls for marking on a single occasion. Very often this design is impracticable because we cannot mark enough individuals at one time to extract an estimate of population size precise enough for our requirements. The alternative calls for marking on several occasions, the population's size being estimated from the rate at which the proportion of marked individuals rises as progressively more are marked.

Schnabel (1938) devised a maximum likelihood estimate appropriate to data of this kind but since her equation must be solved by iteration we will use instead a modified version (Schumacher and Eschmeyer 1943) that yields an explicit solution. It is no less efficient than the maximum likelihood method (Turner 1960b, Phillips and Campbell 1970) and may often be more accurate in that it depends less on random mixing of marked and unmarked individuals (DeLury 1958).

Where N is the constant size of the population, M_i is the number of individuals marked prior to the ith sampling occasion, and n_i is the number of individuals captured on the ith occasion of which m_i had been marked previously, population size can be estimated from

$$
N = \frac{\sum M_i^2 n_i}{\sum M_i m_i}.
$$

The standard error of N is calculated indirectly (DeLury 1958, Ricker 1958: 101). First compute the standard error of $1/N$ as $s/\sqrt{\sum M_i^2 n_i}$ where

$$s^2 = \frac{\sum (m_i^2/n_i) - (\sum M_i m_i)^2/(\sum M_i^2 n_i)}{j-1},$$

j being the number of recapture samples. Confidence limits are then calculated for $1/N$ by multiplying together its standard error and the appropriate value of t corresponding to $j - 1$ degrees of freedom. The confidence limits of N are then calculated by inverting those obtained for $1/N$. Suppose N were estimated as 100 and hence $1/N$ as 0·01. The standard error of $1/N$ is calculated as $\pm 0·001$ from 5 recapture samples. The 95 per cent confidence limits of $1/N$ are therefore $1/N + (0·001 \times 2·776) = 0·0128$ and $1/N - (0·001 \times 2·776) = 0·0072$, the value 2·776 being t at the 0·05 probability level with 4 d.f. By taking the reciprocals of these limits the 95 per cent confidence limits of N are calculated as 78 and 139.

A major advantage of this method lies in the check it allows on the assumption of equal catchability. Unless the assumption is violated the regression of m_i/n_i on M_i is linear through the origin with a slope of $1/N$.

Schumacher's method is more rigidly constrained than Petersen's. It requires that the population maintains a constant size during the experiment and that no animal dies or leaves the area. It shares with the Petersen estimate the assumptions that no animal is born or joins the population during the experiment and that marked and unmarked individuals are equally catchable.

10.5 BAILEY'S TRIPLE CATCH

When animals are born into the population or migrate into the study area between marking and recapturing, neither Petersen's nor Schumacher's method returns an accurate estimate of total numbers. Bailey (1951, 1952) provided one of the several published solutions to this problem. His method yields estimates of birth rate and death rate in addition to population size.

The triple-catch estimate requires data from two marking occasions and two recapturing occasions. The first recapturing and second marking use the same sample so only three catches need be made. Table 10.4 shows the kind of data needed for a triple-catch estimate. Six hundred individuals were captured, marked and released at time 0. At time 1 a sample of 700 animals were caught, the 182 of these bearing marks were released again and the remainder were marked before release. These 518 individuals were given a mark different from the mark used at time 0. The sample captured at time 2 comprised 700 individuals of which 146 had been marked at time 0 and 158 at time 1. Note that no individual carried two marks. Bailey (1951) warned against remarking at time 1 those animals previously marked at time 0. Had this been done, some of them would have been recorded in both the m_{02} and m_{12} recapture samples. Although this procedure is valid statistically it increases any bias resulting from capture-proneness.

The calculations are relatively simple:

Table 10.4.   Data amenable to Bailey's triple-catch analysis

Time	Animals marked	Animals examined for marks	Recaptures from M_0	Recaptures from M_1
0	$M_0 = 600$			
1	$M_1 = 518$	$n_1 = 700$	$m_{01} = 182$	
2		$n_2 = 700$	$m_{02} = 146$	$m_{12} = 158$

$$N_1 = \frac{M_1(n_1 + 1)m_{02}}{(m_{01} + 1)(m_{12} + 1)}$$

$$= \frac{518 \times 701 \times 146}{183 \times 159} = 1822$$

where N_1 is the size of the population at time 1. It has a formal standard error of approximately

$$\text{S.E.} = \sqrt{\left\{ N_1^2 - \frac{M_1^2(n_1 + 1)(n_1 + 2)(m_{02} - 1)m_{02}}{(m_{01} + 1)(m_{01} + 2)(m_{12} + 1)(m_{12} + 2)} \right\}}$$

$$= \sqrt{\left\{ 1822^2 - \frac{518^2 \times 701 \times 702 \times 145 \times 146}{183 \times 184 \times 159 \times 160} \right\}}$$

$$= 238 \text{ (or 13 per cent of } N_1\text{)}.$$

Between times 1 and 2 the rates of birth and immigration combined are estimated as

$$B_{12} = \frac{m_{01}(n_2 + 1)}{n_1(m_{02} + 1)}$$

$$= \frac{182 \times 701}{700 \times 147} = 1 \cdot 240.$$

The combined rates of death and emigration between times 0 and 1 are estimated as

$$D_{01} = 1 - \frac{M_1 m_{02}}{M_0(m_{12} + 1)} = 0 \cdot 207.$$

Since the 0, 1 period is unlikely to be the same length as that of 1, 2 these rates must be reduced to their instantaneous equivalents before much can be done with them:

$$B_{12} = e_{12}^b \text{ to give } b_{12} = 0 \cdot 215$$

and

$$D_{01} = (1 - e^{-d})_{01} \text{ to give } d_{01} = 0 \cdot 233.$$

If the interval 0, 1 contains four weeks and 1, 2 contains six weeks, then on a weekly basis $b_{12} = 0.215/6 = 0.0358$ and $d_{01} = 0.233/4 = 0.0583$. Assuming that these rates have held constant over the ten weeks of the experiment the rate of increase r between times 0 and 2 is

$$r = b - d = 0.0358 - 0.0583 = -0.0225.$$

By extrapolating this rate from time 1 backward to time 0 and forward to time 2 the population numbers at these times can be estimated as

$$N_0 = N_1 e^{-rt} = 1822 e^{-(-0.0225 \times 4)} = 1994 \text{ and}$$
$$N_2 = N_1 e^{rt} = 1822 e^{(-0.0225 \times 6)} = 1592.$$

The triple-catch analysis is specifically designed for cases where individuals are born during the experiment. However, even when population size is measured between birth pulses the method is preferrable to the Petersen estimate because it detects and counteracts the effect of immigration.

Several variants of the triple-catch analysis are available. The papers by Leslie (1952), Leslie and Chitty (1951), Leslie, Chitty and Chitty (1953) MacLeod (1958), Coulson (1962) and Sonleitner and Bateman (1963) provide a lead into the literature, the last also giving a method for estimating the sample size necessary to yield a standard error of specified magnitude.

10.6 THE JOLLY–SEBER METHOD.

When marked animals are recaptured on two or more occasions a stochastic analysis developed independently by Jolly (1965) and Seber (1965) allows an elegant estimation of the parameters of the population. The analysis differs in kind from those of Petersen and Bailey. Theirs treats a probability of dying of, say, 10 per cent as if it removed exactly 10 animals from a population of 100. Their models are 'deterministic'. The Jolly–Seber model on the other hand treats a 10 per cent probability of dying as removing exactly 10 per cent of animals only over a very large number of runs. In the short term it might reduce a population of 100 by 6, 10, 13, 14 or some other number, each of which can itself be assigned a probability. The model is termed 'stochastic'.

The method requires that each animal's history of recapture is known. Either an individual must be marked with a time-specific mark each time it is captured, in which case it may carry several by the end of the experiment, or it must carry a unique mark such as a numbered tag or band. In the second case the number is noted at each capture to build up a history of recapture.

The following notation refers to data collected on the ith occasion:

N_i = estimated population size
n_i = size of sample
M_i = number of marked animals in the population immediately preceding the ith occasion

m_i = number of marked animals in the sample
R_i = number of animals marked and released
r_i = number of animals of the R_i released that are subsequently recaptured,
Z_i = number marked before the ith occasion that were *not* recaptured on on ith occasion but were recaptured subsequently
a_i = proportion of marked animals in the population at the ith sampling.

The Jolly–Seber method is illustrated with data on recapture of the Common Blue Butterfly *Polyommatus icarus* reported by Parr, Gaskell and George (1968). A few imaginary specimens are added to their records to increase low frequencies. A table is drawn up to summarize all information on marking and recapture (Table 10.5). The row labelled *4* on its right shows that $n_i = 8$ animals were recaptured on occasion *4* of which $0 + 2 + 1 = 3$ had previously been marked. Two of these had last been recaptured previously on occasion *2* and one on occasion *3*. The unmarked five animals in the n_i sample were then marked but in the process $n_i - R_i = 1$ animal was killed by mistake. Hence three previously marked and four freshly marked ($R_i = 7$) animals were released. The addition of each column yields r_i, the number of the R_i animals released that were subsequently recaptured. In the example $r_i = 12$ for the sample taken on occasion *2*.

Table 10.6 is now prepared by adding from the left (excluding n_i and R_i) the frequencies in each row of Table 10.5 and entering the cumulative totals, the last of which is parenthesized. Hence the totals in row *6* are, from left to right, 0, $0 + 1$, $0 + 1 + 2$, $0 + 1 + 2 + 1$ and $(0 + 1 + 2 + 1 + 2)$. The parenthesized frequency at the head of each column is m_i, *i* in each case being the italicized number to its right. The entries below that in parentheses in each column are summed to give Z. Hence $m_i = 9$ for occasion *3* and $Z_i = 6$.

Tables 10.5 and 10.6 contain all the information necessary for estimating population parameters. The size of the population at each time of sampling, other than on the first and last occasions, is estimated as

Table 10.5. The n_i, R_i and r_i values for a Jolly–Seber analysis

Number captured n_i	Number released R_i	Time of last previous capture (italicized numbers)					
9	9	*1*					
13	12	5	*2*				
14	14	2	7	*3*			
8	7	0	2	1	*4*		
18	16	1	2	3	1	*5*	
14	14	0	1	2	1	2	*6*
	$r_i =$	8	12	6	2	2	—
		r_1	r_2	r_3	r_4	r_5	

Table 10.6. The values of m_i and Z_i derived from Table 10.5 for a Jolly–Seber analysis

(m_i)	1					
	(5)	2				
	2	(9)	3			
	0	2	(3)	4		
	1	3	6	(7)	5	
	0	1	3	4	(6)	6
$Z_{i+1} =$	3	6	9	4	—	—
	Z_2	Z_3	Z_4	Z_5		

$$N_i = n_i + \frac{n_i Z_i R_i}{m_i r_i}$$

to give

$$N_2 = 13 + \frac{13 \times 3 \times 12}{5 \times 12} = 20 \cdot 8,$$

$$N_3 = 14 + \frac{14 \times 6 \times 14}{9 \times 6} = 35 \cdot 8,$$

$N_4 = 92 \cdot 0$ and $N_5 = 100 \cdot 3$. To each estimate can be attached a formal standard error of

$$\text{S.E.}_{\cdot i} = \sqrt{\left[N_i(N_i - n_i) \frac{M_i - m_i + R_i}{M_i} \left(\frac{1}{r_i} - \frac{1}{R_i} \right) + \frac{1 - a_i}{m_i} \right]}$$

where $a_i = m_i/n_i$ (e.g. $a_4 = 0 \cdot 375$)

and $M_i = m_i + \frac{Z_i R_i}{r_i}$ (e.g. $M_4 = 34 \cdot 5$).

The standard error of N_4, for instance, is

$$\text{S.E.}_{\cdot 4} = \sqrt{\left[92(92 - 8) \frac{34 \cdot 5 - 3 + 7}{34 \cdot 5} \left(\frac{1}{2} - \frac{1}{7} \right) + \frac{1 - 0 \cdot 375}{3} \right]} = 55 \cdot 50.$$

The probability that an individual alive at the moment of release on the ith occasion will survive and not emigrate from the study area before capture of the next sample is

$$p_i = \frac{M_{i+1}}{M_i - m_i + R_i}$$

and so $p_4 = 39/(34 \cdot 5 - 3 + 7) = 1 \cdot 013$ which being greater than unity, and therefore impossible, is interpreted as unity.

The number of animals joining the population by birth or immigration between the ith and $i + 1$ occasion, and which are still alive at the end of this interval, is estimated as

$$A_i = N_{i+1} - p_i(N_i - n_i + R_i)$$

and so

$$A_4 = 100 \cdot 3 - 1 \cdot 013(92 - 8 + 7) = 8 \cdot 12.$$

Jolly (1965) outlined the calculation of standard error for p_i and A_i.

Although this method gives no explicit solution for population size on the first or last sampling occasions these totals can be approximated by a method no less precise than that used in Bailey's triple-catch analysis (Section 10.5). The rate of increase between occasions 2 and 3 ($[\log_e N_3 - \log_e N_2]/t$) is taken as an estimate of rate of increase between occasions 1 and 2. N_1 is estimated by extrapolating backwards at this rate from the N_2 population size. The size of the population at the last sampling occasion is estimated in the same way by extrapolating forward at the rate current between ante-penultimate and penultimate occasions. The Jolly–Seber method may be used as an alternative to Bailey's method when the design calls for only three samplings.

White (1971) provides a FORTRAN listing for computer estimation of population size and other statistics by the Jolly–Seber method.

10.7 ESTIMATION OF MORTALITY RATE

Suppose M animals are marked and released. Subsequently samples obtained with constant effort recaptured m_0, m_1, m_2 and m_3 of these marked individuals which were subsequently released again. These recovery frequencies allow an estimate of mortality rate even when the interval between samplings is variable.

The date on which the m_0 sample is obtained is designated time 0 and time is measured thereafter from that date. A regression of $\log_e m$ on time provides a slope b that is an estimate of $-d$, d being the instantaneous rate of mortality in terms of the chosen units of time. It is converted to the proportion dying over a unit of time by $\bar{q} = 1 - e^{-d}$. Table 10.7 gives an example.

When intervals between recapturing are constant the estimate simplifies to

$$\bar{q} = 1 - \frac{m_1 + m_2 + m_3}{m_0 + m_1 + m_2},$$

the unit of time to which $\bar{q}$ is applicable being the constant interval between recapturing occasions. Instantaneous rate of mortality, d, need be calculated only when $\bar{q}$ must be expressed in different time units. If $\bar{q}$ per week is $0 \cdot 113$ then on a weekly basis

$$d = \log_e \frac{1}{1 - \bar{q}} = 0 \cdot 12$$

and therefore

$$\bar{q} = 1 - e^{-0 \cdot 12/7} = 0 \cdot 017 \text{ per day.}$$

Table 10.7. Calculation of mortality rate from recapture frequencies of $M = 200$ marked individuals

Time (weeks) x	Recaptures m	$\log_e$ recaptures y
0	88	4·477
5	53	3·970
17	10	2·303
21	8	2·079

$\sum x = 43$ $\sum y = 12\cdot829$

$\sum x^2 = 755$ $\sum xy = 102\cdot660$

$$b = \frac{\sum xy - (\sum x)(\sum y)/n}{\sum x^2 - (\sum x)^2/n} = \frac{102\cdot66 - (43 \times 12\cdot829)/4}{755 - (43 \times 43)/4} = -0\cdot120$$

Instantaneous mortality rate $= d = 0\cdot120$ on a weekly basis
Finite rate of mortality $= \bar{q} = 1 - e^{-d} = 0\cdot113$ per week
Finite rate of mortality $= \bar{q} = 1 - e^{-52d} = 0\cdot998$ per year
Finite rate of mortality $= \bar{q} = 1 - e^{-d/7} = 0\cdot017$ per day

This method is often used incorrectly by including in the frequency of recapture for a given occasion not only the marked individuals caught then but also those caught subsequently and which therefore must have been alive on the given occasion even though they escaped capture. This procedure overestimates $\bar{q}$. Another common mistake is to include the number marked, M, in the calculation such that mortality rate is incorrectly estimated as

$$\bar{q} = 1 - \frac{m_0 + m_1 + m_2 + m_3}{M + m_0 + m_1 + m_2}.$$

By this procedure $\bar{q}$ is biased upward.

Mortality rates cannot be estimated from the same data as those used to estimate population numbers from frequency of capture (Section 10.8). The accuracy of such estimates of population size depends on an absence of mortality during the experiments. Experimental designs therefore differ. The experiment to measure the size of a population is run over a very short interval whereas that measuring rate of mortality must continue until a significant proportion of marked animals have died.

10.8 FREQUENCY OF CAPTURE MODELS

Most mark–recapture models are fantasies of the 'what if...?' type, the conjunction being followed by a list of improbable conditions the least practical of which requires that all animals are equally catchable. Recently a good deal of thought has gone into the problem of making mark-recapture models more realistic and hence more likely to return accurate estimates. Most attempts at improvement have been aimed at developing models of frequencies of capture that relax the requirement of equal catchability. As of now little progress can be reported, the analyses published to date being crude and insensitive, but they point the way towards more refined models that will without doubt be developed in the next few years.

Frequency of capture analyses operate on the number of animals caught once, twice, three times and so on over several capturing occasions. These data form a zero-truncated frequency distribution of captures, the missing zero-class representing the unknown number of animals that were never caught. The analysis attempts to estimate the frequency of the zero-class from the shape of the truncated distribution. Population size is then calculated as number of animals captured at least once plus the estimated number that were never caught.

These analyses share a convenient property. Each reveals its applicability to a set of data by the fit of the expected frequencies to the observed frequencies. They need not therefore be used blindly out of faith or desperation. Nor need the analysis be limited to one model, the appropriate strategy being to fit several distributions to the data and to accept that giving the closest fit. The three analyses given here are based on different assumptions about catchability. Between them they cover a wide though not exhaustive spread of the possible ways in which catchability varies. The Poisson method is underpinned by the assumption that catchability is constant whereas the negative binomial and geometric models allow for unequal catchability generated in different ways.

Table 10.8. Zero-truncated Poisson, negative binomial and geometric distributions fitted to capture frequencies of male agamid lizards

Number of captures i	Number of individuals f_i	Poisson $E(f_i)$	Negative binomial $E(f_i)$	Geometric $E(f_i)$
1	13	8·180	5·191	11·017
2	6	8·150	2·244	6·349
3	2	5·413	1·186	3·659
4	2 ⎫	2·697 ⎫	0·684 ⎫	2·108 ⎫
5	1 ⎪	1·075 ⎪	0·414 ⎪	1·215 ⎪
6	0 ⎪	0·357 ⎪	0·259 ⎪	0·700 ⎪
7	1 ⎬ 5	0·102 ⎬ 4·263	0·165 ⎬ 1·842	0·403 ⎬ 4·976
8	0 ⎪	0·025 ⎪	0·107 ⎪	0·233 ⎪
9	1 ⎪	0·006 ⎪	0·071 ⎪	0·134 ⎪
10	0 ⎪	0·001 ⎪	0·047 ⎪	0·077 ⎪
> 10	0 ⎭	0·000 ⎭	0·095 ⎭	0·106 ⎭
χ^2		5·687	24·007	1·128
d.f.		2	1	2
P		0·06	< 0·001	0·5
Estimated N		30	117	45

The use of these models is demonstrated with frequency of capture records of male agamid lizards, *Amphibolurus barbatus* (J. A. Badham, unpublished). Lizards were captured on 21 days over a period of 37 days between birth pulses. Recruitment is therefore zero and the rate of mortality will be relatively low. Table 10.8 shows the frequency with which individual lizards were caught once, twice, three times and so on, the extreme being a lizard caught on 9 of the 21 occasions.

Poisson estimates

A zero-truncated Poisson distribution will fit the observed frequencies closely if catchability is constant. We have met this distribution before in Section 10.11 where it was used to test for equal catchability. Here it is employed to estimate the size of the population. The method was introduced by Craig (1953) and has been used subsequently by Kikkawa (1964) and Eberhardt (1969b).

The fitting of the distribution is described in Section 10.1.1. The mean number of captures per individual caught at least once, $\bar{x}$, is calculated as the mean of the observed frequency distribution of captures, and from it is estimated the mean number of captures, $\bar{X}$, per head of all individuals in the population. In the example (Table 10.8) $\bar{x} = \sum f_i i / \sum f_i = 2.3077$ from which is estimated $\bar{X} = 1.9927$. On the tentative assumptions that catchability was constant and that mortality over the course of the experiment was negligible, population size is estimated as

$$N = \frac{\sum f_i i}{\bar{X}} = \frac{60}{1.9927} = 30.$$

Negative binomial estimates

The negative binomial is a commonly used frequency distribution which often fits the frequency of events whose underlying probability is not constant. For example it usually provides a better fit than does the Poisson to the number of accidents per week that occur in a factory The fit probably reflects variation in accident-proneness between workers. Since accident-proneness and capture-proneness are similar aberrations the negative binomial distribution might be expected also to fit a distribution of capture frequencies. This method of estimating population size was introduced by Tanton (1965) and Holgate (1966).

As with the Poisson method, population size is calculated from the mean frequency $\bar{X}$ with which all individuals in the population are captured. $\bar{X}$ is estimated in turn from the mean frequency of capture $\bar{x}$ of individuals caught at least once. To do so we require a further statistic, s^2, the variance of the frequency of capture of individuals caught at least once. In the example

$$\sum f_i = 13 + 6 + 2 + \ldots = 26,$$

$$\sum f_i i = (13 \times 1) + (6 \times 2) + (2 \times 3) \ldots = 60, \text{ and}$$

$$\sum f_i i^2 = (13 \times 1^2) + (6 \times 2^2) + (2 \times 3^2) \ldots = 242.$$

From these the mean and variance of the zero-truncated distribution of captures is estimated as

$$\bar{x} = \frac{\sum f_i i}{\sum f_i} = \frac{60}{26} = 2 \cdot 3077, \text{ and}$$

$$s^2 = \frac{1}{(\sum f_i) - 1} \left(\sum f_i i^2 - \frac{(\sum f_i i)^2}{\sum f_i} \right) = \frac{1}{25} \left(242 - \frac{60^2}{26} \right) = 4 \cdot 1415.$$

Brass (1958) showed that the mean of the complete distribution can be estimated from

$$\bar{X} = \bar{x} - \frac{s^2 f_1}{\bar{x}(\sum f_i - f_1)}$$

where f_1 is the observed frequency of single captures. Provided the observed distribution is really a zero-truncated negative binomial distribution this method provides an approximate but efficient estimate of $\bar{X}$. In the example

$$\bar{X} = 2 \cdot 3077 - \frac{4 \cdot 1415 \times 13}{2 \cdot 3077(26 - 13)} = 0 \cdot 513.$$

The unknown number of individuals never caught can now be estimated as

$$f_0 = \frac{\sum f_i i - \bar{X} \sum f_i}{\bar{X}} = \frac{60 - (0 \cdot 513 \times 26)}{0 \cdot 513} = 91$$

and the total trappable population as

$$N = \frac{\sum f_i i}{\bar{X}} = \frac{60}{0 \cdot 513} = 117.$$

As a check: $N - f_0 = $ number caught $= \sum f_i = 26$.

The appropriateness of the negative binomial model, and hence the probable accuracy of the estimate of population size, is judged by comparing the observed frequencies with the expected frequencies. To calculate the latter we first compute two additional statistics. Maximum likelihood estimates of these (Sampford 1955) take hours to calculate but Brass (1958) gave approximations that should be adequate for most purposes:

$$w = \frac{\bar{x}}{s^2} \left(1 - \frac{f_1}{\sum f_i} \right) = 0 \cdot 2786$$

$$k = \frac{w\bar{x} - f_1 / \sum f_i}{1 - w} = 0 \cdot 1981$$

The frequencies to be expected if the observed distribution is a zero-truncated negative binomial are calculated as

$$E(f_i) = \sum f_i \times \frac{w^k}{1-w} \times \frac{\Gamma(k+i)}{\Gamma(k)i!} \times (1-w)^i \quad \text{(Fisher 1941)}$$

where $\Gamma(\ )$ means the gamma function, $E(\ )$ means 'expected', ! means factorial and i takes values of 1, 2, 3, ... Gamma functions can be calculated by Sterling's formula (Cramer 1945):

$$\log_e \Gamma(x) = \left(x - \frac{1}{2}\right)\log_e x - x + 0 \cdot 91767 + \frac{1}{12x} - \frac{1}{360x^3}$$

the last term of which can be omitted when x is greater than 3. Expected frequencies calculated by this method are listed in Table 10.8.

Geometric estimates

Eberhardt (1969b) explored the possibility of employing the geometric distribution, a limiting case of the negative binomial distribution. Although he was able to suggest theoretical reasons why the geometric might be expected to fit distributions of frequency of capture, he preferred to introduce the method simply as an empirical result by demonstrating that close fits could be achieved.

The geometric distribution has a single parameter q, the probability that a given individual will be captured at least once. Edwards and Eberhardt's (1967) equation, modified by Chapman and Robson's (1960) correction for bias in small samples, estimates q as

$$q = \frac{\sum f_i i - \sum f_i}{\sum f_i i - 1} = \frac{60 - 26}{59} = 0 \cdot 5763.$$

The size of the population is estimated by $N = \sum f_i / q = 26/0 \cdot 5763 = 45$, and the expected frequencies from $i = 1$ by $E(f_i) = (\sum f_i)(1 - q)q^{i-1}$. These are listed for the example in Table 10.8.

Interpretation of fit

Table 10.8 gives the frequencies to be expected if the observed distribution of capture frequencies was a zero-truncated Poisson, negative binomial and geometric distribution respectively. The goodness of fit in each case is summarized by a χ^2 calculated as $[f_i - E(f_i)]^2 / E(f_i)$ for each i down to $i = 3$, the frequencies being pooled thereafter, and summed over the four classes. The Poisson and geometric values of χ^2 have $n - 2 = 2$ degrees of freedom whereas that of the negative binomial has $n - 3 = 1$.

The probabilities derived from the χ^2 values should not be examined in isolation. Their function in this case is to reveal which of the three models best describes the observed pattern of capture. For the example the geometric distribution is the clear winner and the negative binomial distribution comes a poor last. The looseness of that fit is in large measure a reflection of the approximations used in its calculation — these work well only when the observed distribution closely approximates a negative binomial—but the result still makes the essential point, that the negative binomial model is inappropriate to these data.

The zero-truncated frequency models should not be used independently of each other. They should be used as a triple-ball cartridge. One at least is liable to hit the target. For the data used here the geometric distribution scored whereas the other two were well off. In different field circumstances the Poisson or negative binomial distribution might be more appropriate. Appendix 3 provides a listing of a FORTRAN program for calculating population size and fitted frequencies according to each of the three models.

These methods do not estimate population size if animals die over the course of the experiment. Nor does the fit or lack of fit reveal anything about catchability if the population is open. Tanton (1969) fitted a zero-truncated negative binomial to frequency of capture of two short-lived species of small mammal trapped over $2\frac{1}{2}$ years. His estimate of population size is neither the mean standing population over this period nor the total number of individuals that lived on his study area during this period. Likewise, nothing can be extracted from Bustard's (1969, 1970) demonstration that Poisson distributions did not fit the frequency with which he recaptured two species of lizards over two years. The lizards might well have varied in catchability as he claimed but the failure of fit is not relevant evidence.

10.9 ANALYSIS OF HUNTING RETURNS

The methods discussed previously are workable only when the biologist has complete control over the experiment. More often the design is shaped by conditions over which he has no influence and the experiment must therefore be built around these conditions. This is particularly so when the population is hunted. Marked animals are not recaptured and released again into the population but are killed by hunters who may or may not bother to send the tag or band to the receiving agency. For each tag returned the biologist can determine only when and where the animals was marked and (hopefully) when and where it was killed. He can seldom determine what proportion of the harvest was marked. Although the data are always of this general form the appropriate analysis differs according to whether the hunting season is long or short, whether animals are marked over a short or extended period and whether the marked animals are an unbiased sample of the population or are selected according to age.

Population size is determined by mark–recapture from the ratio of marked to unmarked individuals in a random sample. This statistic can easily be obtained when the experiment is under the control of the experimenter, with more difficulty when the recapturing depends on a commercial fishery or the whaling industry, and with extreme difficulty when recapturing depends on the activity of sportsmen. In theory a marked/unmarked ratio can be estimated by checking a sample of bags, but in practice this information is not forthcoming. Marks returned by hunters can seldom be used to estimate the size of the hunted population. Only mortality rates are derivable from most mark–recapture experiments utilizing hunting returns.

The analyses given subsequently are each based on the assumption that the probability of shooting an animal of a given age is proportional to the frequency of that age class in the population. Although this assumption appears both logical and realistic it is almost the reverse of that underlying the majority of published analyses of game-bird mortality. These usually follow the methods of Bellrose and Chase (1950) which depend on the tacit assumption (contrary to their stated assumption) that the probability of a bird being shot during a hunting season is proportional to the probability of its dying from natural causes over the following year had it not been shot. Hence shooting is considered to sample susceptibility to all causes of mortality, a highly questionable assumption. Neither are the following models constrained by the usual requirement that mortality rate is constant after the first year of life. This hypothesis has not been substantiated for any vertebrate group; nor is it a mathematical necessity of analysis.

Whatever the conditions of the experiment, we can usually estimate the proportion of the total population, or at least a sex-age segment of it, that dies from all causes over a year. If sportsmen returned all the tags they recovered this proportion could be partitioned into those animals killed and recovered during the hunting season and those that died from other causes during the year. Included in the second category are those animals shot but not recovered and those that were recovered but the tags not returned. Sportsmen do not return all tags. Rates of return have variously been estimated in the United states as about 30 per cent (Crissey 1967), 40 per cent (Tomlinson 1968), 45 per cent (Bellrose 1955), and 50 per cent (Geis and Atwood 1961), but with considerable variation between states. Estimates of this kind provide correction factors allowing an estimate of the rate of hunting mortality.

Isolated rates of hunting mortality and natural mortality can be calculated only when the hunting season is short and the animals are tagged a short time before the season opens.

The following methods are designed to cope with the range of situations faced by a wildlife manager. Alternative approaches are provided by Seber (1970, 1972) and Johnson (1974).

General assumptions and notation

The following assumptions underly all the analyses given in this section. Additional assumptions, specific to a particular model, are outlined when that model is introduced.

1. Once marked, animals do not lose their marks, or if they do the effect of loss is corrected.

2. Hunting pressure does not steadily increase or decrease with time although it may fluctuate from year to year.
3. Only those marks recovered by hunting are included in the analysis.
4. All ages are susceptible equally to hunting.

The following notation is common to all models:

M = number of individuals marked and released.

m = number of marks recovered during a hunting season.

x = age in years, scaled by an amount specific to each model. In the first, for example, age $x = 0$ is equivalent to an age of six months and hence $x = 1$ is equivalent to $1\frac{1}{2}$ years.

u = proportion of individuals above base age (i.e. $x = 0$) killed by hunting over a year.

w = proportion of individuals above base age killed by other agents of mortality over a year.

h = isolated annual rate of hunting mortality.

n = isolated annual rate of other losses.

q = annual rate of total mortality, the proportion of individual dying from all causes during a year.

When these rates are specific to a given interval of age they are subscripted according to the scaled age at the beginning of the interval. Thus q_x is the rate of total mortality between scaled ages x and $x + 1$. A bar over the symbol indicates a mean annual rate for all age above base age, weighted by the representation of each age class in the populations. The various mortality rates are related by $q = u + w$ and $q = h + n - hn$.

10.9.1 Analyses appropriate when only juveniles are marked

Model 1: Animals are marked in only one year and recovered over several years, the hunting season is short, the animals are marked shortly before it commences, and hunting pressure is the same from year to year.

Table 10.9 gives Balham and Miers' (1959) records of bands returned from 1069 juvenile black ducks, *Anas superciliosa*, banded in 1950. Base age ($x = 0$) is the mean age of juveniles at their first hunting season, about half a year in this example.

First calculate the rate of total mortality for each age interval as

$$q_x = 1 - \frac{m_{x+1}}{m_x} = 1 - \frac{12}{18} = 0.333 \text{ for age } x = 2.$$

That for age 4 is not calculated because the frequencies of recapture are too low to return a meaningful estimate, and q_x for age 5 is impossible to calculate from these data. The weighted mean annual mortality rate is calculated as

Table 10.9. Recovery frequencies and mortality rates of male black ducks banded as juveniles in a single year (data from Balham and Miers 1959)

Scaled age	x	0	1	2	3	4	5
Recoveries	m_x	314	69	18	12	2	1
Initial number	M	1069					
Partitioned rates of mortality $\begin{cases} \bar{u} \overset{?}{=} h_x \\ w_x \\ n_x \end{cases}$		0·294	0·294	0·294	0·294		
		0·486	0·445	0·039	0·539		
		0·688	0·630	0·055	0·763		
Rate of total mortality	q_x	0·780	0·739	0·333	0·833		

$$\bar{q} = \frac{m_0}{\sum m_x} = \frac{314}{416} = 0 \cdot 755.$$

The proportion killed by hunting over one year can be estimated only for the age interval 0, 1. We assume that it is constant over all age intervals at a level of

$$\bar{u} = \frac{m_0}{M} = \frac{314}{1069} = 0 \cdot 294.$$

The proportion of an age class dying from other causes over a year is then

$$w_x = q_x - \bar{u} = 0 \cdot 333 - 0 \cdot 294 = 0 \cdot 039 \text{ for age } x = 2.$$

If we are willing to assume further that the mortality between banding and the onset of the first hunting season was negligible, the isolated rates of hunting mortality and natural mortality can be calculated respectively as $h_x = \bar{u}$ and

$$n_x = \frac{q_x - h_x}{1 - h_x} = \frac{0 \cdot 039}{0 \cdot 706} = 0 \cdot 055 \text{ for age } x = 2.$$

Table 10.9 lists age-specific mortality estimates calculated in this way.

Model 2: Juveniles are marked shortly before a restricted hunting season in each of several years and are recovered by hunting over several years.

When several cohorts can be pooled to estimate mortality rates, the constancy of hunting pressure demanded by the first model is here no longer imperative. Calculations are similar to those of model 1 except that values of M are now specific to age.

Table 10.10 gives a set of data of this kind. The zero frequencies are real; dashes indicate classes for which no entry is possible. For example, no frequency can be entered for the recovery of yearling or older birds banded in year 7 because the experiment was terminated after the hunting season of that year. Likewise, no frequency of recovery is available for birds older than one year that were banded as juveniles in year 6 or for birds older than two years banded

Table 10.10. Recovery frequencies and mortality rates of male black ducks banded as juveniles in each of several years (data from Balham and Miers 1959)

Year banded	Number banded	Recoveries (m) by age (x)							Annual rate of hunting mortality
		0	1	2	3	4	5	6	
1	175	64	15	4	1	0	0	1	0·366
2	1069	314	69	18	12	2	1	—	0·294
3	452	112	28	12	4	0	—	—	0·248
4	330	92	11	10	4	—	—	—	0·279
5	312	80	17	10	—	—	—	—	0·256
6	124	17	9	—	—	—	—	—	0·137
7	27	8	—	—	—	—	—	—	0·296
Total	2489								
Recoveries m_x		687	149	54	21	2	1	1	
		of	of	of	of	of	of	of	
Initial number M_x		2489	2462	2338	2026	1696	1244	175	
Partitioned rates of mortality $\begin{cases} \bar{u}=h_x \\ w_x \\ n_x \end{cases}$		0·276 0·505 0·698	0·276 0·342 0·472	0·276 0·275 0·380	0·276 0·610 0·843				
Rate of total mortality q_x		0·781	0·618	0·551	0·886	$\bar{q}=0·731$			

in year 5, and so on. The effect of this progressive truncation must be corrected when cohorts are pooled to estimate age-specific mortality rates. Since M is no longer a constant, age-specific values of M_x are calculated as the number of birds originally banded that could have contributed to the m_x recoveries. M_x is calculated by subtracting from the total banded the number banded in the years for which a dash is entered in the column of age-specific recoveries (Table 10.10). The initial size of the 'cohort' providing 54 recoveries of two-year-old birds is $M_2 = 2489 - 27 - 124 = 2338$.

Mortality rates are estimated by equations analogous to those of model 1:

$$q_x = 1 - \frac{M_x m_{x+1}}{M_{x+1} m_x} = 1 - \frac{2338 \times 21}{2026 \times 54} = 0·551 \text{ for age 2};$$

$$\bar{q} = \frac{m_0}{M_0 \sum (m_x/M_x)} = \frac{687}{2489 \times 0·3777} = 0·731;$$

$$\bar{u} = \frac{m_0}{M_0} = \frac{687}{2489} = 0·276;$$

$$h_x = \bar{u} = 0·276;$$

$$w_x = q_x - \bar{u} = 0·551 - 0·276 = 0·275 \text{ for age 2; and}$$

$$n_x = \frac{q_x - h_x}{1 - h_x} = \frac{0.551 - 0.276}{0.724} = 0.380 \text{ for age 2.}$$

The estimates are checked at each age by

$$q_x = \bar{u} + w_x = h_x + n_x - h_x n_x.$$

Proportion killed by hunting each year, $\bar{u}$, can be calculated only for the first hunting season after banding. Its calculation presupposes that sportsmen return all recovered bands. Since this assumption will be wrong to a greater or lesser degree $\bar{u}$ is biased downward and w_x is therefore biased upward, although q_x is not affected. These biases are always present unless m_0 is corrected for the bands collected but not returned. The bias of $\bar{u}$ has little effect on the usefulness of $\bar{u}$ as a measure of hunting pressure and it can confidently be employed to monitor year to year fluctuations in the intensity of hunting. The last column of Table 10.10 gives annual estimates of $\bar{u}$. In year 4, for example, the rate of hunting mortality was $92/330 = 0.279$. These estimates vary considerably from year to year, thereby vindicating the pooling of several cohorts to smooth out differences between years. However since they do not show a time trend general assumption 2 is not violated.

Model 3: Animals marked as juveniles shortly before an extended hunting season in each of several years are recovered by hunting over several years.

This model relaxes the requirement that the hunting season is short but in so doing sacrifices the capacity of estimating h_x and n_x. In this sense it is inferior to model 2 and should be used only when the assumptions of that model are entirely unrealistic. The tacit assumption of model 2—negligible mortality during the hunting season other than that caused by hunting—must be rejected if the season spans more than a month. But when most of the hunting is concentrated at the beginning of the season the effect is the same as if the season were short; the data may be analysed by model 2. In the examples used to illustrate the first two models the season for black ducks varied between two and four weeks but Balham and Miers (1959) showed that 64 per cent of the harvest was taken in the first two days of each season.

The difference between this model and the last can be appreciated most easily if we interpret the data in Table 10.10 as if it were collected during hunting seasons lasting six months. Cohorts are pooled in the same way as for model 2 to give the same m_x and M_x frequencies. The difference in treatment lies in reinterpreting the scaled age x and in omitting calculation of h_x and n_x. Isolated rates of hunting mortality, h_x, cannot be calculated because the number of recoveries during the hunting season is now a function of both hunting pressure and the rate of mortality from other causes during the season. These cannot be separated. For the same reason n_x is also lost. In the previous models ages were scaled such that $x = 0$ represented an age of about six months. In this model age $x = 0$ is the mean age of juveniles at the middle of the hunting season. If the season were six months long and started at the same date as the short seasons considered previously, $x = 0$ is equivalent to about $\frac{3}{4}$ of a year of age and $x = 1$ is $1\frac{3}{4}$ years. Mean mortality rate $\bar{q}$ is therefore the proportion of those birds older than nine months that die from all causes during a year.

Model 4: Marking and recovery over several years; significant mortality between marking and the first hunting season.

The analysis is the same as for model 3 whether the hunting season is restricted or extended. Since the banded cohort is reduced by natural mortality before entering its first hunting season, h_x and n_x cannot be calculated for any age.

The base age $x = 0$ is the mean age of juveniles at the middle of their first hunting season.

10.9.2 Analyses appropriate when animals of all ages are marked

In most large-scale marking operations all captured animals are marked. The marked sample at the moment of release is therefore close to a random sample of the population's age distribution and sex ratio at that time. Returns one year after marking are no longer from a random sample because all the marked individuals are now at least a year old. Returns at two years after marking are from animals that are two years of age or older. Hence the ages of animals shot 0, 1, 2, 3 and so on years after marking are drawn from a series of age distributions progressively truncated from the lower end.

Model 5: animals are marked at random with respect to age a short time before a restricted hunting season.

Table 10.11 lists the frequency with which bands were returned from male black ducks banded at all ages (Balham and Miers 1959). Since the birds are considered a random sample of the population's age distribution at the time of marking, the frequencies in the body of the table no longer relate to age x but rather to years after banding, i.

Mortality rates calculated from these data are not age-specific but are means for all age classes combined, each class contributing to the mean in proportion to the number of animals of that age in the population as a whole. Four frequencies only are needed for analysis: M_0, M_1, m_0, m_1. The other frequencies are redundant.

A comparison of Table 10.10 (analysis of black ducks banded as juveniles) and Table 10.11 (analysis of black ducks banded irrespective of age) can produce information not revealed by the two tables in isolation. Such a comparison is often possible because many banding operations return both sets of data.

Firstly we compare rates of hunting mortality. That for juveniles was $\bar{u} = 0.276$ as against $\bar{u} = 0.267$ for all ages. We deduce from their closeness that a juvenile and an adult have the same chance of being killed during the hunting season, that vulnerability to hunting is not a function of age. Secondly we compare the schedule of w_x for birds banded as juveniles with $\bar{w}$ for all ages to find whether

Table 10.11. Recovery frequencies and mortality rates of male black ducks banded without regard to age in each of several years (data from Balham and Miers 1959)

Year banded	Number banded	Recoveries (m) by years since banding (i)							Annual rate of hunting mortality
		0	1	2	3	4	5	6	
1	212	71	17	4	1	1	0	1	0·335
2	1419	395	90	30	19	5	2	—	0·278
3	590	146	35	18	5	1	—	—	0·247
4	411	112	14	14	4	—	—	—	0·273
5	358	89	21	10	—	—	—	—	0·249
6	138	20	9	—	—	—	—	—	0·154
7	27	8	—	—	—	—	—	—	0·296
Total	3155								
Recoveries m_i		841	186	76	29	7	2	1	
Initial number M_i		of 3155	of 3128	of 2990	of 2632	of 2221	of 1631	of 212	

Partitioned rates of mortality

$$\left\{\begin{array}{l} \bar{u} \overset{?}{=} \bar{h} = \dfrac{m_0}{M_0} = \dfrac{841}{3155} = 0\cdot267 \\[2mm] \bar{w} = \bar{q} - \bar{u} = 0\cdot777 - 0\cdot267 = 0\cdot510 \\[2mm] \bar{n} = \dfrac{q - \bar{h}}{1 - \bar{h}} = 0\cdot696 \end{array}\right.$$

Rate of total mortality
$$\bar{q} = 1 - \frac{M_0 m_1}{M_1 m_0} = 1 - \frac{3155 \times 186}{3128 \times 841} = 0\cdot777$$

the isolated rate of mortality from causes other than hunting is related to age. We find that the rate for all birds over a year ($\bar{w} = 0\cdot510$) is largely a reflection of the rate of first-year mortality ($w_0 = 0\cdot505$) and that the rate at subsequent years of age declines and then rises again. Natural mortality rates apparently vary with age. Thirdly we compare $\bar{q} = 0\cdot731$ of birds banded as juveniles with $\bar{q} = 0\cdot777$ for birds banded irrespective of age. These are two estimates of the same parameter: the proportion of the total population dying each year. The difference here probably reflects sampling variation. Had it been larger we would suspect either that the birds banded irrespective of age were not a random sample of the population's age distribution or that rate of increase diverged sharply from zero. This comparison acts as a control for interpreting differences between the other rates. Before a difference can be taken seriously it must be larger than the difference between the two estimates of $\bar{q}$.

A more critical comparison can be made between juveniles and adults rather than between juveniles and all ages. Balham and Miers (1959) presented the necessary data but its analysis does not alter the conclusions given above.

The analysis outlined in Table 10.11 is seldom used. More often a rate of total mortality is calculated for each year after banding and these are averaged to give a 'mean annual mortality rate'. Let us examine this procedure. Total mortality rate over the first year after banding represents the proportion of birds of all ages that die during a year from all causes. The rate over the second year after banding is an estimate for birds of all ages other than those in their first year. That for the third year after banding is applicable to all birds except those in their first and second years. A 'mean annual mortality rate' calculated by averaging these successive estimates is heavily weighted towards older birds. Weighting is fine so long as the contribution of each age class to the mean is weighted according to the proportion of birds of this age in the population ($\bar{q}$ of Table 10.11 is such an estimate) but the process of averaging mortality rates of successive years after banding has precisely the opposite effect; the weights are inversely proportional to age-class frequencies. Thus the mortality rate of three-year old birds is given three time the weight of juvenile mortality although juveniles outnumber them by around ten to one in most populations of game birds. Unless the rate of mortality is constant with age this procedure will return an inaccurate estimate of the proportion of the population that dies each year.

Model 6 : Animals of all ages are marked in several years before an extended hunting season.

Mortality rates are calculated as for model 5, but because $\bar{h}$ is no longer estimated by $\bar{u}$ neither $\bar{h}$ nor $\bar{n}$ can be extracted from the banding returns. Total mortality rate $\bar{q}$ is interpreted as the proportion of the total population dying from all causes between the middle of one hunting season and the middle of the next. This estimate is a weighted mean over all age classes other than the class of animals younger than the interval between the middle of the season of hatching and the middle of the hunting season.

Model 7 : Hunting and marking continue throughout the year.

Since animals are being marked over the same period as they are being hunted the data collected in this way differ in kind from those discussed previously. Two courses are open. If the date of marking and of death is known for each animal whose band was returned the data can be grouped into intervals of time between marking and death, irrespective of the season of these two events, and analysed as for model 6. This method ignores seasonal variation in mortality and the results must therefore be interpreted cautiously. Alternatively, if we need to consider mortality in terms of the cycle of seasons, the data must be grouped by year of death, the year being defined arbitrarily as between, say, January and December or between July and June. In this way we keep together deaths that occurred in a season when mortality is high. Should most animals die in the period December to February, for example, the year would be defined to start in, say, July.

The second alternative runs us into a further problem. The number of bands returned in the same year as they were attached cannot be interpreted as for the previous models. A band attached at the beginning of the year has a far higher probability of return during that year than one attached in the last month of the year. Consequently, frequency of return during the year of banding is not comparable to frequency of return in subsequent years. We have two choices: either to estimate mortality in the year of banding by making assumptions about rate of marking and rate of mortality throughout the year, or to reject returns of animals shot in the year they were marked and to estimate rate of mortality from the remaining records. Ricker (1948: 62, 1958 : 118) gave equations for estimating $\bar{q}$ and $\bar{u}$ from recoveries in the year of marking and the subsequent year. Their validity depends on the rates of marking and of hunting being constant throughout the year. Beverton and Holt (1957 : 199) showed that a solution was theoretically possible even when these rates varied by season but their analysis requires additional information that is seldom available.

In the above analysis (Table 10.12) of Frith's (1963) banding records of grey teal, *Anas gibberifrons*, in Australia, recoveries during the year of banding are rejected because the information needed to correct them is not obtainable. Frith arbitrarily dated the beginning of the year as 1 July so that the southern summer, when he expected most mortality, would not be split between years.

The last column of Table 10.12 is again calculated by dividing returns in the year of banding by the number banded in a given year. It cannot now be interpreted as the proportion of the population shot during a year but it serves as

Table 10.12. Recovery frequencies and mortality rate of grey teal banded throughout the year and hunted throughout the year (data from Frith 1963)

Year banded	Number banded	Recoveries (m) by years since banding (i)						Index of annual rate of hunting mortality
		0	1	2	3	4	5	
1954–55	21	1	1	0	0	0	0	0·048
1955–56	291	26	15	4	0	2	1	0·089
1956–57	1724	201	71	26	11	3	2	0·117
1957–58	11295	618	168	131	77	34	9	0·055
1958–59	1802	174	63	48	12	6	—	0·097
1959–60	254	14	5	7	0	—	—	0·055
1960–61	536	27	10	5	—	—	—	0·050
Total	15923							
Recoveries m_i		1061	333	221	100	45	4	
		of	of	of	of	of	of	
Initial numbers M_i		15923	15923	15923	15387	15133	13331	

$$\bar{q} = 1 - \frac{M_1 m_2}{M_2 m_1} = 1 - \frac{15923 \times 221}{15923 \times 333} = 0·336$$

a check on the general assumption that hunting pressure did not regress on time during the experiment. The values are erratic but show no obvious time trend.

Only $\bar{q}$ can be estimated from these data. It is interpreted as the proportion of the population above a certain age that died from all causes between 1 July and 30 June. The 'certain age' depends on whether the population has a birth-pulse or a birth-flow breeding system, and if it is the former, the interval between the birth pulse and the arbitrary date defining the beginning of the year. Grey teal can be treated as a birth-flow population in Australia if band returns from several years are pooled (Frith 1959). The base age of $\bar{q}$ in Table 10.12 is therefore the mean age of juveniles at banding ($\frac{1}{2}$ year) plus one year. No additional information can be extracted from these records.

Chapter 11

Population analysis in management

The aims of population management are few and specific, and they can usually be achieved by specific action. There are, in fact, only three problems of population management:

1. the treatment of a small or declining population to raise its density,
2. the exploitation of a population to take from it a sustained yield, and
3. the treatment of a population that is too dense, or which has an unacceptably high rate of increase, to stabilize or to reduce its density.

These three problems are labelled respectively as conservation, sustained yield harvesting, and control. Some fields of management are concerned only with one or two of these problems. Management of fish populations is aimed almost exclusively at harvesting a sustained yield. The other two problems seldom arise. The management of insect populations, in contrast, is oriented towards control. Wildlife management, which will be discussed here at greatest length, deals with all three problems. In every case the problem is solved by a manipulation of the dynamics of a population, and in this sense population analysis is central to its solution, but the means by which the manipulation is effected will vary greatly according to circumstances. Sometimes a modification of available shelter, food or water supply will trigger the desired change in the population's rate of increase. Sometimes a more direct manipulation may be called for.

Although there are but three problems of population management, these are not the source of the myriad arguments over what should be done where, when. No problem is a problem of itself. A thing or a relationship or a situation becomes a problem only when it does not suit our purposes, whatever those purposes might be. The arguments usually revolve around the appropriate purpose of a particular piece of land and its human and non-human inhabitants. No management decision is possible until we know what it is that we want.

Only then can we decide whether a harvesting scheme is appropriate or whether the population conflicts with our stated aims by being too sparse or too dense. If we intended to convert a piece of land into a national park we would view with pride and approval a population of large herbivores living in the area. Our eyes might narrow if we envisaged the same area as a potential market garden. A large proportion of the effort expended on population management, particularly wildlife management, is devoted to finding out whether a problem exists and if so whether it can be solved by harvesting, conservation or control. Population analysis has nothing to do with this question. It is a technique that may be extremely useful in solving a problem but it is useless until that problem has been defined.

11.1 CONSERVATION

Many problems of conservation turn out to be, on close examination, not technical problems but problems of economics and sociology. In many instances no great insight is needed to discover why a population is declining, and counter-measures, if not immediately practicable, are at least theoretically specifiable. The main difficulty lies in persuading people to take appropriate action. This difficulty is exemplified by the exploitation of the humpback whale, *Megaptera novaeangliae*. Chittleborough (1965) summarized the history of the population that summers in the Southern Ocean below Western Australia (the group IV population of whaling literature). Between 1912 and 1934 only a few of its members were killed each year in the Antarctic and along the migration route up the west coast of Australia. Hunting intensified in the following five years, at least 12,700 whales being taken during commercial whaling operations. Towards the end of this period the steep decline in catch per unit effort signalled a rapid depletion of the stock. Then came the war. For ten years few whales were taken, the population increasing again under this regime of *de facto* protection. When catching resumed in 1949 whaling was controlled in Antarctic waters by the International Whaling Commission, first by a quota system and then by a time limit (generally four days) on the season. Quotas were set for the Australian shore stations by the Australian government. In 1960 the International Whaling Commission, alarmed at the decline in humpbacks, proposed that hunting of the group IV population be suspended in Antarctic waters for three years. The governments of Japan, Norway, U.S.S.R. and U.K. rejected the recommendation and catching continued. The quotas set for the shore stations by the Australian government were reduced progressively from 1959 but the yearly trend in quota and catch (Table 1 of Chittleborough 1965) makes little sense unless the quotas, far from being estimates sustained yield, were actually estimates of the maximum number of whales the shore stations could secure when working at full capacity. This uncharitable interpretation may be false but I doubt it. Under the sutained onslaught of pelagic and shore-based whaling the population collapsed in 1962. The decline was sudden and spectacular: within a year the population declined to such a size that it could no longer support commercial operations, either shore based or pelagic, relying solely on humpbacks. Chittleborough estimated that, under complete protection, the population would take about 50 years to recover.

In one way this is a good example of a conservation problem that was really an economic problem. Wisdom is often confused with hindsight, but in this instance the decisions made for economic reasons were known to be ecologically wrong by the people who made them at the time they made them. No technical problem was involved. The dynamics of that population were understood in some detail. The mass of technical data collected on that population could only be interpreted in one way: the population was declining and the decline was a direct consequence of excessive hunting.

But in another way the decline of the group IV humpback population is atypical of most problems of conservation. Very few populations are now declining because they are hunted; destruction and modification of food supply and habitat are presently the major threats to vertebrate populations. At an outright guess I would estimate that 95 per cent of conservation problems are in this category. Even where hunting appears to be the main reason for the decline of a population, closer investigation usually implicated induced changes in food, shelter, or water supply.

The technical part of tackling a conservation problem—discovering why a population is declining and deciding what can be done about it—can range from very easy to very difficult. Since there is little point in launching a five-year study of the problem if the answer might be found in the first week, the investigation is best carried out in two stages. If an answer is not forthcoming from the simple investigation of the first stage a more detailed investigation is begun.

Initial investigation

A study of the reasons why a population is declining is made immeasurably more simple if we have access to a population of the same species that is in healthy condition. Habitat, food supply, water supply, fat reserves and incidence of disease of the threatened population can then be compared with those of the healthy control population. This often shows up a difference that can be investigated further. If, for example, the threatened population lived in an area grazed by sheep but sheep were absent from the range of the control population, and if this were the only difference we could detect between the environments of the two populations, we would hypothesize that sheep in some way caused the decline. The conclusion would be tested by taking off the sheep and noting whether the decline slowed and then reversed.

A large proportion of conservation problems can be cracked by a simple investigation of this kind. The method carries its own safety device in that the result of the treatment prescribed to halt the decline is the test of the hypothesis explaining the cause of the decline.

Some conservation problems are not solvable by a stage 1 investigation and we must therefore dig deeper.

Stage 2 investigation

At this stage we already know from the abortive result of the initial investigation that the cause of the decline is neither simple nor obvious. We will change our tactics accordingly from assault to siege.

Firstly we measure fecundity rates to find whether or not the decline is a consequence of breeding impairment. If it is we must search further for the cause. Breeding impairment is an uncommon cause of decline except in some populations of predatory birds. Its discovery would suggest that breeding physiology is being affected either by disease or the presence of a metabolic poison in the environment.

Since a decline can result only from either reduced fecundity or increased mortality, if fecundity rates are found to be at healthy levels the problem is unambiguously one of excessive mortality. Should mortality be excessively high at the adult stage but juvenile mortality is at a normal level for the species this is *prima facie* evidence that the decline is caused by excessive hunting. Most other agents of mortality act more heavily on juveniles. Should the decline be a consequence of reduced food supply or poor habitat the diagnostic feature is a rise in juvenile mortality, usually coupled with a decline in juvenile fecundity.

The process of identifying the segment of the population that is in trouble can be speeded up if a life table is available for the threatened population. A comparison of this table with one from a healthy population will reveal immediately which age-class is being affected. Getting a life table from a threatened population requires ingenuity—obviously we are not going to shoot a sample—but it can be done. We might, for example, calculate a distribution of ages at death, S'_x, from a picked-up sample of skulls and convert this to a schedule of d_x by $d_x = S'_x e^{rx}$, r being the rate of decline and therefore negative (Section 8.1.2, method 5).

Having isolated the age classes in trouble, either from increased mortality or lowered fecundity, the isolation of causes is simply a routine investigation.

Where decline in the quality of habitat is suspected as the cause of the decline in numbers a simple check is to enquire of the local people whether they have noticed any change in density of vegetation, depth of swamps and so on over the last ten years. A more precise check is provided by aerial photographs. Few parts of the world have escaped being photographed from the air. Many regions have been covered at least twice, thereby allowing an assessment of the trend in habitat between the two surveys, and between the last survey and the present. I have seen aerial photographs taken ten years apart that required considerable study to convince me that they were of the same area, such was the change in habitat wrought by fire, cattle grazing and agricultural encroachment.

11.2 PRINCIPLES OF HARVESTING

To most people the harvesting of populations connotes whaling, sealing and commercial fishing, activities carried out by professionals for profit. But in terms of men and time these activities are only a minor branch of population harvesting; most harvesting effort is expended by sportsmen.

Whether a population is managed for commercial harvesting or for recreational hunting or fishing, the same principles apply. Management is aimed at providing a sustained yield (SY), a crop that can be taken year after year without forcing the population into decline. An SY is not a unique value for a given population. There will be a large number of SY values each corresponding to a different management treatment. Efficient management aims at developing the treatment to provide the largest possible sustained yield (the maximum sustained yield, MSY) or more commonly to provide a sustained yield that maximizes revenue or recreational values (the optimum sustained yield, OSY).

The rate of harvesting that holds rate of increase at $r = 0$ is that rate at which a population would increase if it were not harvested. Hence a population increasing at a finite rate of $e^r = 0.20$, and therefore at an exponential rate of $r = 0.182$, can be harvested at an instantaneous rate of $H = 0.182$. The instantaneous rate of harvesting requires some explanation. It is the rate at which the population could be harvested throughout the year. Symbolizing as $\bar{N}$ the mean number of individuals in a population throughout the year, the sustained yield per year totals $H\bar{N}$ animals. But when the population is harvested during only part of the year, the SY is calculated not from the instantaneous rate of harvesting but from the isolated rate of harvesting, h, where $h = 1 - e^{-H}$. A numerical example will make this clear. Suppose a birth-pulse population increasing at $r = 0.182$ contains 1000 animals immediately after the birth pulse, and that the severity of natural mortality is constant throughout the year at a rate of $\bar{n} = 0.25$ per year (equivalent to 0.0236 per month). Numbers in successive months after the birth pulse will conform to the series for the unharvested population in Table 11.1. By the end of the year the population is reduced by mortality to 750 individuals which give birth to $m = 0.6$ offspring per head, upping numbers to $750 + 750\,m = 1200$ at the birth pulse.

We will now harvest this paper population to hold its rate of increase at $r = 0$ by allowing sportsmen to harvest it over a short hunting season. To calculate the appropriate rate of harvesting, and hence the number of animals that the hunters will be allowed to kill, the instantaneous rate of harvesting, $H = 0.182$, is converted to the isolated rate of hunting mortality, h, by

$$h = 1 - e^{-H} = 0.167.$$

At any time of the year 0.167 of the animals then alive can be harvested over a short period of time to hold rate of increase to zero. Obviously the appropriate SY will differ according to when the hunting season is declared, because natural mortality progressively reduces between birth pulses the number of animals

Table 11.1. Effect on numbers of harvesting a birth-pulse population once during the year (see text)

Month	Unharvested numbers	Harvest in first month	Harvest in fifth month
1	1000	$1000 \xrightarrow{h=0.167} 833$	1000
2	976	813	976
3	953	794	953
4	930	775	930
5	908	757	$908 \xrightarrow{h=0.167} 757$
6	887	739	739
7	866	722	722
8	846	705	705
9	826	688	688
10	806	672	672
11	787	656	656
12	768	640	640
1	$750 + 750m$ $= 1200$	$625 + 625m$ $= 1000$	$625 + 625m$ $= 1000.$
	$m = 0.6$ $r = 0.182$ SY $= 0$	$m = 0.6$ $r = 0.0$ SY $= 167$	$m = 0.6$ $r = 0.0$ SY $= 151$

exposed to risk of death by hunting. Table 11.1 shows the effect of harvesting immediately after the birth pulse when 167 (1000×0.167) can be shot, and of harvesting in the 5th month when 151 (908×0.167) can be shot. Under either harvesting regime the numbers at the next birth pulse are held to 1000 and the population's rate of increase is therefore held to zero.

Should we wish to harvest twice during the year, the appropriate rate of harvesting at each occasion is calculated by halving H and again calculating an isolated rate of hunting mortality:

$$_2h = 1 - e^{-H/2} = 0.087.$$

Table 11.2 shows the effect of harvesting at this rate on two occasions, firstly in the 9th and 11th months and secondly in the 6th and 7th. The final result is the same: population size at the next birth pulse is held to that at the last birth pulse, i.e. $r = 0$.

By extension, the appropriate rate of harvesting can be calculated for any number of occasions during the year. For example, if we harvested in 12 separate, equally-spaced occasions, the rate of harvesting appropriate to any one occasions would be

$$_{12}h = 1 - e^{-H/12} = 0.015.$$

It results in an SY of 146 animals per year. But if animals are to be harvested with

Table 11.2.   Effect on numbers of harvesting a birth-pulse population on two occasions during the year (see text)

Month	Unharvested numbers	Harvests in 9th and 11th months	Harvests in 6th and 7th months
1	1000	1000	1000
2	976	976	976
3	953	953	953
4	930	930	930
5	908	908	908
6	887	887	$887 \xrightarrow{2h=0\cdot087} 810$
7	866	866	$722 \xleftarrow{2h=0\cdot087} 791$
8	846	846	705
9	826	$826 \xrightarrow{2h=0\cdot087} 754$	688
10	806	736	672
11	787	$656 \xleftarrow{2h=0\cdot087} 719$	656
12	768	640	640
1	$750 + 750m = 1200$	$625 + 625m = 1000$	$625 + 625m = 1000$
	$m = 0\cdot6$ $r = 0\cdot182$ SY $= 0$	$m = 0\cdot6$ $r = 0\cdot0$ SY $= 134$	$m = 0\cdot6$ $r = 0\cdot0$ SY $= 146$

this frequency they might as well be harvested throughout the year. The effect is about the same. With continuous harvesting at the appropriate rate the population's average size throughout the year is $\bar{N} = 800$, and the yield is therefore

$$SY = H\bar{N} = 0\cdot182 \times 800 = 146 \text{ animals.}$$

These numerical examples illustrate general rules of harvesting:

1. An SY is calculated from the instantaneous rate of harvesting, H, which equals the rate of increase, r_p, the population would assume if it were not harvested.

2. When the year is divided into short intervals the population may be harvested during any one of these at a rate $h = 1 - e^{-H}$ of animals surviving at that time.

3. When harvesting is spread over more than one interval the appropriate rate of harvesting during each is $1 - e^{-H/y}$, where y is the number of short harvesting intervals.

4. An SY accrues from harvesting in any short interval at any rate providing that these rates expressed in instantaneous form jointly total the overall instantaneous rate H, where $H = r_p$, the rate at which the population would increase if it were not harvested.

5. As a close approximation, the above four rules hold whether rate of natural mortality is constant or variable throughout the year and whether the population has a birth-pulse, birth-flow or intermediate breeding system, providing that these aspects of the life history remain constant from year to year.

The numerical examples dealt with an imaginary population which, in the absence of harvesting, had a positive r. We harvested it mathematically to reduce its rate of increase to zero and to hold its numbers constant at the level they had reached at the beginning of the harvesting program. It is an unlikely example because most populations, although fluctuating from year to year, have a rate of increase which, averaged over several years, is close to zero (see, for instance, the trend in the size of the deer population graphed in Figure 4.3). If the rate of harvesting depended solely on a population's rate of increase, as it does in the previous examples, a population with an average rate of increase of $r = 0$ has an appropriate instantaneous harvesting rate of $H = 0$ and therefore an SY of zero.

A population of constant size can be harvested only after it has been manipulated to raise r above zero. Harvesting management makes use of the fact that, in general, the greater is the availability of food and other resources per head, the greater is the population's rate of increase. A positive rate of increase is therefore generated by lowering density, thereby increasing the resources available to the survivors. If left alone after the reduction the population would eventually climb back to its previous density and resume a mean r of zero. But if, after the initial reduction, the population were harvested at the same rate as it seeks to increase, the density would then be stabilized by the harvesting.

We would expect that, within limits, the farther a population is reduced below its unharvested density the higher will be its induced rate of increase; and the higher will be the rate of harvesting needed to hold numbers stable at the reduced density. Thus for each density there is a potential rate of increase, r_p, that would be manifested if harvesting were stopped (Section 5.3). Put in another way, for each density there is a corresponding harvesting rate of $H = r_p$ that holds the population stable in size. In terms of actual yield rather than rate of harvesting, for any given density we can estimate the number of animals that must be harvested each year to stabilize the population. Although H tends to increase as density is lowered the SY does not increase in parallel. H is small when density is reduced only a little and the appropriate SY is a small fraction of a comparatively large population. When density is reduced substantially the population, which is now comparatively small, will have a high r_p and hence a high H. The appropriate SY is now a large fraction of a small population. The maximum sustained yield (MSY) of a population harvested unselectively with respect to sex and age is taken from the population size N where HN is at its maximum. The task of estimating the MSY is therefore a complex dual operation. We must calculate both the density from which

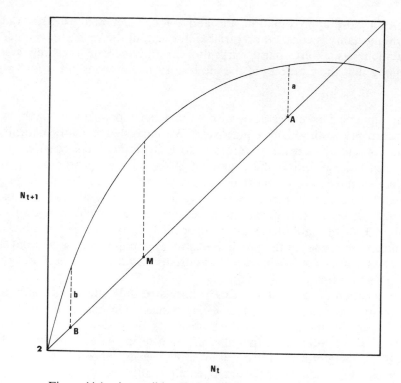

Figure 11.1 A possible relationship (curve) between numbers in one year and numbers the year after. The length of the dashed lines is proportional to the sustained yield at different values of N_t (adapted from Ricker 1958).

an MSY can be harvested and the rate of harvesting that stabilizes density at that level. That sturdy work-horse of wildlife management, the technique of successive approximations, is unequal to this task. The probability of successfully approximating to the appropriate value of HN by varying H and N independently is low enough to make the search for the proverbial needle in the haystack a certain bet by comparison.

Figure 11.1 illustrates these points. It represents an idealized relationship between numbers (or density) in one year and numbers a year later. The diagonal line gives all possible positions for $N_{t+1} = N_t$, i.e. when numbers do not change between two successive years. A point falling to the left of the diagonal represents an increase of numbers between times t and $t + 1$; a point to the right records a decline. The curve in the same figure describes a relationship that might be expected between a population's size in two successive years when the supply of food is replenished at a constant rate. Its shape will differ between species and between populations within species. The level of N_t where the curve cuts the diagonal is the 'steady density', the average size of the unharvested population.

No population is as neat as this. The environment fluctuates from year to year causing fluctuations in survival, fecundity, age distribution and rate of increase, and the dependence of present numbers on numbers the year before is therefore weak. The graph represents a relationship that might well underly the fluctuations of a simple population in a simple environment. We would expect the trend to be blurred by environmental and sampling effects but not to be obliterated.

Figure 11.1 shows three strategies whereby this idealized population could be harvested for sustained yield. In the first the population is reduced from its steady density to a size of $N_t = A$. The dashed line a, vertically linking the population curve and the diagonal, represents the number that would be added to the population over the next year. This is proportional to the number that must be harvested each year to hold the population stable at size $N_t = A$. The SY is a small proportion of a large population. Alternatively, the population could be reduced to $N_t = B$ and an SY proportional to the length of line b could be harvested each year, the yield being a large proportion of a small population. The MSY is taken at the population size where the diagonal and the population curve have the maximum vertical separation, at $N_t = M$.

Figure 11.1, for all its abstraction, illustrates several points applicable to harvesting real populations:

1. If a population is stable in numbers it must be reduced below its steady density before it can be harvested.

2. For each density to which a population is reduced there is an appropriate SY.

3. For each level of SY there are *two* levels of density from which this SY can be harvested.

4. At only one density can an MSY be harvested.

5. The SY taken each year must be less than the number taken in the initial reduction.

6. When a constant number is harvested from the population each year, the population will decline to and stabilize at the upper size for which that number is the sustained yield. Should the number exceed the MSY the population will decline to extinction.

7. The SY must be calculated from the size to which the population is initially reduced, not from the steady size.

The last is important. The question: what percentage of the population can be harvested each year? is meaningless. It must be rephrased as: if the population is reduced by 20 per cent (or 30 per cent or 40 per cent) what fraction of the reduced population must be harvested each year to hold it stable at that size? Or better still both unknowns can be combined in one question: to what size must the population be reduced, and what fraction must be harvested from it each year thereafter, to provide the maximum sustained yield?

11.2.1 Estimating harvests by the logistic model

If a mean trend of N_{t+1} on N_t could be estimated accurately from field data, the calculation of both the MSY and the population size appropriate to it would be a simple technical exercise. Unfortunately a plot of these pairs usually comprises a considerable scatter of points delineating no clear, unambiguous trend. We have two choices: either to accumulate so many data that the trend of N_{t+1} on N_t is finally signalled clearly through the noise, or to fit a curve to what data are available. These alternatives do not exclude each other. A population can be harvested according to a trial estimate of MSY while further data are collected to provide a more refined estimate. Although we would be more satisfied with a trend deduced directly from the data, the paucity and variability of the information usually forces us to assume the general form of the relationship before estimating its parameters from the data. Because this procedure carries with it an inbuilt potential for error, the effect of the calculated SY must be monitored to determine whether, in fact, the harvesting regime stabilizes the population at the calculated density.

There are a number of models of population growth to chose from. Beverton and Holt (1957 : 327) reviewed those relevant to sustained-yield harvesting up to 1953 and presented several of their own. Subsequently, Andrewartha and Birch (1954: Ch. 14), Ricker 1954a and b, 1958), Leslie (1959), Gulland (1962, 1968) and Lefkovitch (1966, 1967) contributed new models or offered modifications of previous models.

Of those available the logistic will be considered first and in greatest detail. It is not the best model, being oversimple compared with the complexity of real populations, but its very simplicity makes it an ideal introduction to the general strategy of estimating sustained yields. Armed with that understanding a population manager will be in a better position to adapt a model to the special features of the population he seeks to harvest and of the environment to which it reacts.

The logistic model can be used directly for some harvesting problems. Whaling biologists have found that it provides relatively accurate predictions, and experimental manipulations of laboratory populations (Silliman and Gutsell 1958) indicate that is applicable to the dynamics of at least some fish species.

The basic assumption of the logistic model is that r_p and hence H are inversely proportional to density. By extension, the addition of each new recruit reduces by a constant amount the survival or fecundity, or both, of all members of the population. We would not expect so nice a relationship in practice but we hope, in using the model, that something of the sort happens. To be more realistic we would expect that the shortage of resources occasioned by the arrival of an additional recruit would affect younger animals more than older animals. Ricker (1954a, 1958) constructed a model to mirror such an effect, and Murphy (1966) gave a clear summary of the logic underlying it. The main assumption of Ricker's model is almost the reverse of that underpinning the logistic model

(Frank 1960): adult fecundity and survival is not influenced by the addition of a recruit but the survival of the recruit is dependent on the number of adults in the population. One might expect that Ricker's model could result in a markedly different pattern of growth from that generated under logistic assumptions, and so it can. Murphy (1967) has shown, however, than when r_m is below about 0·34 per year the difference is slight. At $r_m = 0·2$ the difference is insignificant.

These findings encourage a different view of the logistic model. Since the dynamics of vertebrate populations are likely to fall somewhere between those defined respectively by the logistic model and by Ricker's model, and since these two models result in a similar outcome of population growth unless r_m is high, the logistic model might be expected to mimic the growth of a population whose dynamics differ from those implied by the logistic equation. It is with this rationale that the logistic model is used here. We need not necessarily assume that population processes and logistic assumptions are congruent when a logistic model is used to estimate sustained yield.

As derived in Section 9.9.1, logistic growth may be symbolized by

$$r = r_m(1 - N/K)$$

where
$N = $ population size,
$r = $ its exponential rate of increase,
$r_m = $ its intrinsic rate of increase, and
$K = $ population size at steady density.

Although models similar to the logistic had been used previously to estimate SY (see the models of Baranov 1926 and of Hjort, Jahn and Ottestad, 1933, for instance), Graham (1935) was first to apply the logistic equation to sustained-yield. He showed that because rN is the instantaneous production of animals surplus to holding the population stable it is also the sustained yield. When H is defined as the instantaneous rate of harvesting necessary to hold N constant, the equation can be rewritten as

$$HN = r_m N(1 - N/K)$$

or as

$$HN = r_m N - \frac{r_m}{K} N^2.$$

The second version has the general form $y = bx - cx^2$ which is the algebraic definition of an upwardly convex parabola that passes through the origin. Since $SY = HN$ for any level of N, such a relationship between population size and SY leads to a very simple calculation of maximum sustained yield: the MSY is harvested from a population size of $N = K/2$ at an instantaneous rate of $H = r_m/2$ to yield $HN = r_m K/4$ animals each year. Put more simply, the population is reduced to half its steady density and harvested at half its intrinsic rate of increase. Figure 11.2 diagrams the relationship between logistic growth and sustained yield.

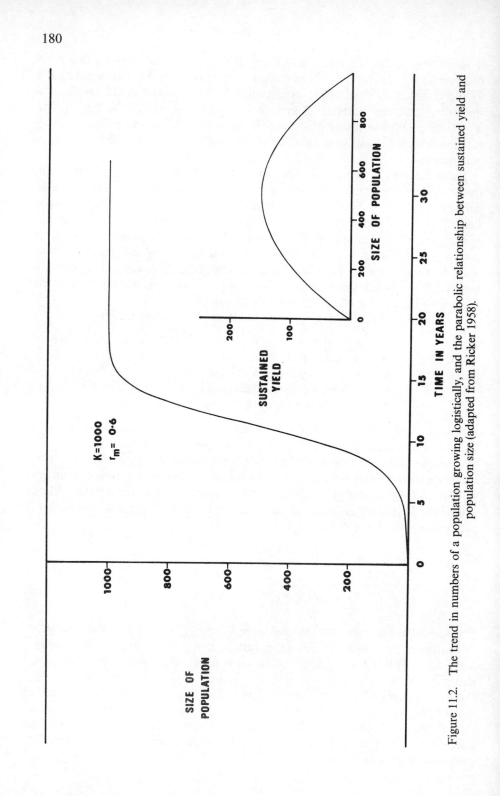

Figure 11.2. The trend in numbers of a population growing logistically, and the parabolic relationship between sustained yield and population size (adapted from Ricker 1958).

If we are willing to assume that the logistic model provides an adequate description of a population's pattern of growth, estimates of K and r_m are sufficient to allow estimation of the MSY. How we go about estimating MSY depends on whether the population is currently being harvested, and whether or not it is currently at steady density.

Population is increasing; neither K nor r_m are known

If a population is increasing towards an assumed steady density of unknown magnitude a logistic curve can be fitted to the trend of numbers with time to provide estimates of r_m and K. Morisita (1965) showed that during logistic growth the trend of $(N_{t+1} - N_t)/N_t$ is linear on N_{t+1} such that

$$(N_{t+1} - N_t)/N_t = a - bN_{t+1}$$

where $a = e^{r_m} - 1$ and $b = a/K$. Consequently both r_m and K can be estimated by least-squares regression if estimates of N are available from three or more consecutive years. MSY can then be estimated from the instantaneous rate of harvesting $H = r_m/2$ appropriate to a population of size $K/2$.

Table 11.3. Calculation of maximum sustained yield for a population with known steady size of $K = 1000$ animals

Year	Numbers (N)	$\dfrac{K - N}{N}$	$\log_e \left(\dfrac{K - N}{N}\right)$
(t)			(y)
0	800	0·250	− 1·39
1	850	0·176	− 1·74
2	890	0·123	− 2·10

$$-r_m = \frac{\sum ty - (\sum t)(\sum y)/n}{\sum t^2 - (\sum t)^2/n}$$

$K = 1000$ $n = 3$

$\sum t = 3$ $\sum y = -5·23$

$\sum t^2 = 5$ $\sum ty = -5·94$

$\sum t^2 - (\sum t)^2/n = 2$ $\sum ty - (\sum t)(\sum y)/n = -0·710$

$$-r_m = -0·710/2 = -0·355; \quad r_m = 0·355$$

$H = r_m/2 = 0·177$ at $N = K/2 = 500$

$$\text{MSY} = 0·177 \times 500 = 86$$

Population is increasing; K known but r_m unknown

If a population has been reduced below its known steady size K, r_m and hence MSY can be estimated from the rate at which it climbs back towards K. If growth is logistic,

$$\log_e\left(\frac{K-N}{N}\right) = a - r_m t,$$

r_m being the slope coefficient of the regression of $\log_e (K-N)/N$ on time t. Table 11.3 demonstrates the calculation for an imaginary population that is reduced from $K = 1000$ to 800 animals. The following year the population has increased to 850 and the year after that to 890. The MSY of 86 animals is calculated, for purposes of illustration, directly from the instantaneous rate of harvesting, H, as it would be if the population were harvested at the same rate throughout the year. Had the population been harvested over a short season the MSY would have been calculated from the finite rate h.

K and one SY are known

Suppose a population has been harvested for some years by taking an SY of known size from a population of known size. Suppose further that the size of this population was also known before the harvesting program was started. We have therefore a known value of K, an SY per year, and the size of the population from which this SY is taken. From these data we calculate the intrinsic rate of increase as

$$r_m = \frac{KH_1}{K - N_1}$$

where H_1 is the instantaneous rate of harvesting that yields the known SY at the known population size N_1. The MSY is therefore harvested from a population of size $K/2$ at an instantaneous rate of $H = r_m/2$.

SY is known for two levels of harvesting

Suppose the SY is known for two intensities of harvesting but no measure of K is available. The two SY values might have been obtained, for instance, if a population had been hunted at a constant rate for many years, the population size adjusting to this rate, and then hunting effort was intensified to shift the size of the population to a second level at equilibrium with a higher yield. Such a set of data might reflect either good luck or good management.

We need to calculate both K and r_m to estimate the MSY and the population size from which it can be harvested. The following example is imaginary but the parameter values are about right for a population of white-tailed deer. For ten years the population had been harvested at a rate that produced a yield averaging 384 deer per year from a population averaging 4,800 animals at the beginning of each hunting season. The game department then decided that the number of hunting licences could be doubled. Under this new regime of harvesting the yield per year finally stabilized at an average of 546, taken from a population averaging 3,900 at the beginning of each hunting season. Since the hunting season is short and most deer are shot during the first few days of the season, the rates of harvesting calculated as yield/stock are isolated rates of hunting mortality, h. From these the instantaneous rates of harvesting are calculated as $H = -\log_e(1 - h)$. The data are:

	Harvesting level (1)	Harvesting level (2)
$N =$	4,800	3,900
$SY =$	384	546
$h = SY/N =$	0·080	0·140
$H = -\log_e(1 - h) =$	0·083	0·151

The intrinsic rate of increase is related to stock and yield at each level of harvesting by

$$r_m = \frac{KH_1}{K - N_1} = \frac{KH_2}{K - N_2}$$

and this equality can be manipulated to provide an estimate of K:

$$K = \frac{H_1 N_2 - H_2 N_1}{H_1 - H_2}$$

$$= \frac{(0·083 \times 3900) - (0·151 \times 4800)}{0·083 - 0·151}$$

$$= 5,900 \text{ deer.}$$

K can now be fed into either of the equations relating r_m to H and N:

$$r_m = \frac{KH}{K - N}$$

Level (1)

$$\frac{5900 \times 0·083}{5900 - 4800} = 0·45,$$

Level (2)

$$\frac{5900 \times 0·151}{5900 - 3900} = 0·45.$$

From these estimates of K and r_m the population size appropriate to the MSY is calculated as $K/2 = 2,950$ which is harvested at an instantaneous rate of $r_m/2 = 0·22$. The appropriate rate of harvesting over a short hunting season is therefore $h = 1 - e^{-0·22} = 0·20$ to give an MSY of $0·20 \times 2950 = 590$, a significant improvement on the previous two levels of sustained yield.

Population is unharvested and neither r_m nor K are known

All too often we are faced with a complete lack of data on a population which, for economic or political reasons, is to be harvested without delay. This is not the tragedy it might appear. If the population is harvested in a systematic and controlled manner, and if the trend in numbers is monitored over the first few years, the MSY and the population size for which it is appropriate can be estimated from the data provided by the harvesting program. The harvesting itself is used as an experimental treatment that reveals the statistics of population growth, and from these the MSY can be estimated.

Since the population is unharvested at the beginning of the program the numbers at that time are taken as an estimate of K. It can be estimated by any of the census techniques outlined in Section 4.2 but perhaps the most appropriate is the index-manipulation-index method (third part of Section 4.2.5) because, in any event, the population is going to be manipulated by the harvesting and we might as well use this to our advantage. By this method population size is calculated from the difference between two indices of abundance measured before and after the first harvest. Suppose that the first density index is calculated as $I_1 = 145.3$. After $C = -75$ animals are harvested from the population the density index is again calculated to give $I_2 = 134.4$. The size of the population before the harvest is therefore

$$K = \frac{I_1 C}{I_2 - I_1} = \frac{145.3 \times (-75)}{134.4 - 145.3} = 1000 \text{ animals.}$$

Note that C is negative because animals were removed from the population, not added to it.

Should we continue harvesting 75 animals each year, the population will eventually decline to and stabilize at the greater of the two densities for which 75 is the SY. The intrinsic rate of increase is estimated from the rate of this decline. At least three estimates of population size, taken immediately before the harvest in successive years, are needed for this calculation. Each is approximately related to that of the year before by

$$N_{t+1} = N_t(1 - h)e^{(r_m - bN_t)}$$

where b is the constant amount by which r_m is reduced by the addition of one recruit and the other symbols are as previously defined. By substituting C/N_t for h and rearranging the equation, we arrive at

$$\log_e\left(\frac{N_t + 1}{N_t - C}\right) = r_m - bN_t,$$

which allows a solving of r_m as the y-intercept of a linear regression. Table 11.4 outlines the calculation for a model population whose parameters of growth are close to those that might be expected from a population of game birds.

The solution of r_m provided by this method is an approximation whose accuracy increases with decrease in the length of the harvesting season. Formally, the yield can be varied each year throughout the course of the experiment, but for practical reasons it is best held constant from year to year.

We now have estimates of K and r_m allowing an estimate of MSY in the

Table 11.4. Estimating r_m from the decline in numbers caused by harvesting $C = 75$ birds per year

N_t (x)	N_{t+1}	$\dfrac{N_{t+1}}{N_t - C}$	$\log_e\left(\dfrac{N_{t+1}}{N_t - C}\right)$ (y)
925	893	1·050	0·0488
893	877	1·072	0·0695
877	869	1·084	0·0808
869	865	1·089	0·0854
865	863	1·092	0·0880

$$\bar{x} = 885\cdot8 \qquad\qquad \bar{y} = 0\cdot0745$$

$$\sum x = 4429 \qquad\qquad \sum y = 0\cdot3725$$

$$\sum x^2 = 3{,}925{,}589 \qquad\qquad \sum xy = 328\cdot398$$

$$\sum x^2 = (\sum x)^2/n = 2380\cdot8 \qquad \sum xy - (\sum x)(\sum y)/n = -1\cdot5628$$

$$b = -1\cdot5628/2380\cdot8 = -0\cdot000656$$

$$r_m = \bar{y} - b\bar{x} = 0\cdot0745 + (0\cdot000656 \times 885\cdot8) = 0\cdot66$$

same way as previously. In the example $K = 1000$ and $r_m = 0\cdot66$, implying that the MSY is harvested at $H = 0\cdot33$ from a population of 500. The MSY taken during a short season is therefore $500h$, where $h = 1 - e^{-H}$: an annual yield of 140 birds.

11.2.2 Estimating harvests by the interactive model

Section 9.9.2 introduced an interactive model for the growth of a herbivore population. Its pattern of growth is an outcome of the dynamic characteristics of the herbivore interacting with the dynamics of the vegetation on which it feeds. To this extent it is more realistic than the logistic model.

In section 9.9.3 the model was elaborated to include a predator, and the equations for the standing crop of plants and animals at equilibrium with the pressure of predation were given for the special case where that predator is a man armed with a high-powered rifle. This model is structured around eight constants of population growth and the rate of grazing. If these could be estimated for a vegetation–herbivore system the MSY of herbivores, and the population size from which it comes, could be calculated with fair accuracy. At this stage of the game that would be too much to hope for. Eventually, however, the managers of terrestrial populations will join fisheries biologists in worrying about the realism of their models and in doing something about it.

There are approximate methods for calculating MSY by the interactive model which circumvent the need to estimate the full range of parameters. Although approximations, they are liable to return estimates more accurate than logistic estimates. These calculations are published elsewhere (Caughley 1976), but it is worth noting here that they lead to estimates of the MSY generally lower than the equivalent logistic estimates. Further, the MSY comes from a larger population ($0.7K$) than that estimated by logistic methods. The disparity between the two sets of estimates transmits yet another warning against an uncritical choice of models for estimating MSY.

11.2.3 Estimating MSY from yield and effort

When fecundity and mortality rates react without delay to a change in food supply or some other resource, the change being faithfully mirrored by an alteration in the population's rate of increase, MSY can be estimated from the relationship between yield and effort.

Figure 11.3 illustrates such an analysis for yellowfin tuna. Schaefer's (1957) data on catch between 1934 and 1955 are plotted against effort. In the same way as the regression of SY on density is parabolic when the population grows logistically, SY also traces a parabola when regressed against the effort expended to obtain that yield. The parabola fitted to the data of Figure 11.3 peaks at 18·6 million lbs (8.5×10^6 kg), the estimated MSY. Schaefer used a slightly different analysis but reached substantially the same conclusion.

Because this method is so dependent on an absence of lag between cause (a lowering of density by harvesting) and effect (attainment of the r_p appropriate to the new density) it should be used only on those populations that turn over rapidly. Populations of large mammals, which at any given time comprise several over-lapped generations, are buffered to some extent against the effects of changing density. The method is not suited to these populations but it might be worth trying on game birds and small mammals.

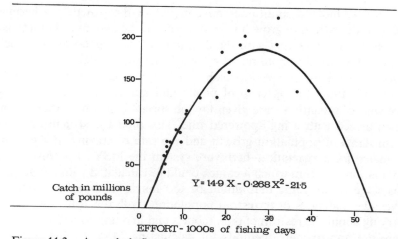

Figure 11.3. A parabola fitted to the relationship between catch of yellowfin tuna and the effort expended, to estimate the maximum sustained yield (data from Schaefer 1957).

This technique requires accurate information on yield and effort, and its efficacy is increased if the biologist holds some control over hunting intensity. He can then increase the rate of harvesting in steps separated by periods in which the rate is held constant. Only the harvest obtained the year before each quantal change in effort should be plotted on the graph of yield against effort. It is this harvest, taken at the end of a period over which hunting intensity is held constant, that will provide the closest approach to a yield at equilibrium with annual effort.

Points on a regression of yield against effort should be accumulated until the regression starts to bend. A parabola of the form $y = a + bx + cx^2$ is then fitted by the method of least squares (Snedecor and Cochran 1967 : 453) to these points and the height of the parabola is taken as an estimate of the MSY if the constant a is close to zero.

11.2.4 Harvesting social groups

Two schemes of harvesting are available when the population is made up of social groups. Either some groups can be harvested in their entirety, leaving other groups wholly intact, or a few individuals can be harvested from each group. The choice of tactics depends on the social organization of the groups and the ranging behaviour of the species.

The strategy of harvesting is aimed at increasing resources per head to enhance rate of increase. The tactics must reflect this overall aim. In general, when group ranges overlap or when the species is nomadic, harvesting should be directed at whole groups, not individuals. Disturbance is minimized thereby and the food and other resources made available by the removal of a group are open to utilization by other groups. When group ranges do not overlap and when dispersal into unoccupied but favourable areas is slow, a few individuals should be harvested from each group. No benefit accrues from increasing resources per head unless the additional resources are used.

11.2.5 Selective harvesting

So far we have dealt only with unselective harvesting, the probability of harvesting an individual being independent of its sex and age. Although inefficient this method is often the only practical procedure. Small mammals trapped for their fur and game birds lacking sexual dimorphism cannot be harvested selectively. Nor can fish, other than through the weak selectivity imposed by mesh size or a length limit.

The commercial cropping of large terrestrial mammals allows greater scope for selective harvesting, as does the recreational hunting of some game populations. Since several sex-age categories can be recognized in the field, unselective harvesting is neither necessary nor desirable. The problem of apportioning harvesting effort most efficiently between segments of the population differs according to whether we select according to sex or according to age. No either-

or decision is implied here—ultimately the harvesting rate of each sex-age class should be optimized—but for the sake of clarity these interlocking problems will first be considered as if they posed two separate questions.

Selective harvesting of age classes

The principles underlying selective harvesting can best be introduced by again considering how unselective harvesting effects the age distribution of females in a birth-pulse population. We have already seen (Section 9.5) that it does not change a population's age distribution from that prevailing before a harvesting program is initiated. The stable age distribution, S_x, of the unharvested population is related to survival and rate of increase by $S_x = l_x e^{-rx}$. Under a regime of unselective SY harvesting age-specific survival is lowered to l'_x such that $l'_x = l_x e^{-rx}$. The effect of unselective SY harvesting is therefore to change the l_x schedule and decrease r to zero so that the survival schedule under harvesting, l'_x, is the same as the unharvested age distribution $l_x e^{-rx}$. The question at issue is whether this is the best manipulation, whether the l_x schedule can be manipulated in other ways without upsetting the requirement that harvesting maintains r at zero. In fact, the *only* constraint on harvesting for sustained yield is that $\sum l'_x m_x = 1$. The balance can be achieved by a wide variety of harvesting tactics generating a multiplicity of l'_x schedules.

Section 8.3 outlined how the mortality rate, q_x, over a given age interval x, $x+1$ can be partitioned into isolated rates of harvesting mortality, h_x, and of natural mortality, n_x, by $q_x = h_x + n_x - h_x n_x$, or in terms of the equivalent survival rates, $p_x = (1 - h_x)(1 - n_x)$. Consequently $\sum l'_x m_x = 1$, the equation symbolizing the constraint that $r = 0$ under SY harvesting, may also be written as

$$\sum_{1}^{\infty} m_x \prod_{y=0}^{x-1} p_y = 1$$

(see Table 8.2 for conversion of l_x to p_x) and therefore as

$$\sum_{1}^{\infty} m_x \prod_{y=0}^{x-1} (1 - h_y)(1 - n_y) = 1.$$

In this form the equation defines the legitimacy of a schedule of age-specific harvesting rates and allows us to test whether a given selective harvesting strategy will stabilize r at zero. Known schedules of n_x and m_x are combined with the desired schedule of age-specific harvesting rates, h_x, and substituted into the equation. If it does not solve as unity the rate of increase under this harvesting strategy will differ from the required value of zero. But this is not enough. We are not interested primarily in whether a given strategy of harvesting holds the population constant in size; we wish to know which strategy of the many that achieve this provides the maximum sustained yield. It is a complex problem. The age distribution of the yield is determined by the schedule of age-specific harvesting rates in use, but so also is the age distribution of the population to be harvested in the following year. The optimal harvesting strategy is a compromise between taking as many females as possible from those ages that contribute least to the population's potential rate of increase, and at the same time setting up the age distribution of the survivors to allow enough females to enter these age classes to provide the yield in the following year.

These conflicting objectives tend to cancel each other. Except in rather special cases there is usually little or nothing to be gained from harvesting age classes at different rates. The optimum strategy turns out to be the unselective harvest. If at a given density the potential rate of increase of the female segment of the population is, say, $r_p = 0.2$, no manipulation of age-specific harvesting rates will increase r_p permanently beyond 0.2 while density is held constant. The exceptions to this rule are trivial. Firstly, when the fecundity rate of females decreases steeply after a given age, harvesting of these older females at a higher rate will increase the MSY above that gained by unselective harvesting. I do not know of a single vertebrate species whose fecundity declines steeply enough with age to make economic sense of harvesting older females at a higher rate. Secondly, if the removal of a young animal greatly increases the viability of the next offspring produced by its mother, a higher rate of harvesting may be justified for the youngest age class. The gain in yield per unit effort, however, is unlikely to be other than marginal. Thirdly, when the depression of r_p associated with the presence of a given female is independent of her weight, and when growth is rapid and asymptotic weight is high, selective harvesting of older females will raise the MSY measured in units of carcase weight. Again I do not know of a vertebrate species where these properties are sufficiently developed to make selective harvesting economically attractive. For practical purposes, then, unselective harvesting is no less efficient than harvesting selective of age, and since it costs less it is the more appropriate strategy.

These conclusions have been stated baldly to avoid losing them amongst a covey of equations. Age-selective harvesting is not a useful strategy because the mutual dependence of the age-specific rates is almost complete. Suppose the female segment of a population is harvested for the MSY. The year classes 0, 1, 2 and $> 2 = \infty$ can be recognized in the field and could therefore be harvested at different rates. Under any scheme of harvesting the stable age distribution at the birth pulse can be written:

1	(at age 0)
$(1 - h_0)(1 - n_0)$	(at age 1)
$(1 - h_0)(1 - n_0)(1 - h_1)(1 - n_1)$	(at age 2)
$\dfrac{(1 - h_0)(1 - n_0)(1 - h_1)(1 - n_1)(1 - h_2)(1 - n_2)}{h_\alpha + n_\alpha - h_\alpha n_\alpha}$	(at ages ∞)

with the constraint that

$$\sum_{1}^{\infty} m_x \prod_{y=0}^{x-1} (1 - h_y)(1 - n_y) = 1,$$

an equality indicating that these harvesting rates stabilize the size of the population. When the stable age distribution is written in this extended form it reveals that the stable frequency at each age contains a term for the rate of harvesting of each younger age class. By virtue of the relationship expressed by the constraining equation, the stable frequency at age zero is a function of the rate at which all age classes are harvested. Such an interlocking system conforms to the truism of physics that the behaviour of an isolated system is unaffected by a rearrangement of its internal relationships. Since we measure the behaviour of a harvested population in terms of the MSY it provides, we can conclude that the yield will not be improved much by changing from unselective to age-selective harvesting. An obvious corollary suggests itself—that when a population is harvested efficiently the imposition of a minimum size limit will have little or no effect on the MSY.

Although these conclusions were drawn specifically for females, they also apply, at least in a general way, to males. It is most unlikely that age-specific harvesting will increase the yield of males beyond that provided by the unselective MSY.

When MSY is measured in terms of value rather than as numbers cropped, the strictures against age-specific selective harvesting are no longer valid. Any combination of yield and effort can also be expressed as revenue and cost. When animals of a given age have a higher market value than those of other ages the MSY is not now the greatest number than can be taken but the combination of ages taken as a sustained yield that returns the highest net revenue. The appropriate age-specific rates of harvesting for one sex are those maximizing the expression

$$W_v h_0 (1-n)^v + \sum_{x=1}^{\infty} W_{x+v} h_x (1-n_x)^v \prod_{y=0}^{x-1} (1-h_y)(1-n_y),$$

given that

$$\sum_{1}^{\infty} m_x \prod_{y=0}^{x-1} (1-h_y)(1-n_y) = 1,$$

where v is the fraction of the year represented by the time between the birth pulse and the harvest and W_x is a weighting factor expressing the relative benefit of harvesting an animal aged x, and subscripts denote age in years. Maximization of this expression by hand is completely impracticable but it is a comparatively simple job to program a computer for that analysis.

Table 11.5 lists the schedules of n_x and m_x estimated from the female segment of a thar population increasing at approximately $r = 0.12$. For simplicity the population is assumed to be at half its steady density and to contain 1000 females at the birth-pulse. The three sets of weighting coefficients are respectively those appropriate when (a) the yield is maximized in terms of numbers, (b) in terms of carcase weight, and (c) in terms of value. The column of coefficients labelled (b) are average age-specific weights in kg. Those labelled (c) are arbitrary in this example, but for a real harvesting program they would represent the

Table 11.5. Maximization of yield from female thar. The population contains 1000 females at the birth pulse and is harvested six months after the birth pulse. See text

Age (yr) x	Natural mortality rate n_x	Fecundity rate m_x	Weighting coefficients W_{x+v}		
			(a)	(b)	(c)
0	0·37	0·00	1	9	2
1	0·03	0·00	1	25	10
2	0·04	0·33	1	32	5
$>2 = \alpha$	0·16	0·45	1	40	1

Yield maximized in terms of:	Harvesting rates				Relative yield		
	h_0	h_1	h_2	h_α	numbers	kg	$
(a) numbers	0·460	0·000	0·000	0·000	**100**	900	200
(b) weight	0·000	0·000	0·000	0·203	52	**2095**	52
(c) value	0·000	0·460	0·000	0·000	72	1800	**720**
unselective	0·113	0·113	0·113	0·113	100	2062	398

net value realized by one female of that age. These may decline with age when skins are harvested (e.g. with caracul sheep and seals) or may rise with age when meat is harvested. But not necessarily so. If the harvested population lives in rough country a young animal may return a greater net profit than an older and heavier one. A hunter may be able to carry a small carcase to a track and transport it to the road by pack horse, whereas a larger carcase must usually be recovered by helicopter. Net profit is higher for the smaller carcase because of reduced overheads.

The lower half of Table 11.5 lists the results of a computer analysis of the optimum rates of harvesting and the resultant yields when the population is harvested to maximize yield in terms of numbers, carcase weight, and value, respectively. The analysis recommended that only the first year class should be harvested if we wish to maximize numbers, but the yield is much the same as for unselective harvesting. This is not an anomaly; it indicates only that there is no unique summit on the maximizing surface. In practice this recommended harvesting schedule would be rejected because age-selective harvesting provides no greater yield and is more expensive to implement than unselective harvesting. The recommendation for maximizing yield of carcase weight would also be rejected because the small gain over unselective harvesting is not enough to justify the added expense it would entail. Only when yield is tallied in terms of value does age-selecting harvesting provide a clear economic gain over unselective harvesting. However the arbitrary coefficients of value used in this example are more variable than would usually be the case. For most harvesting programs the value of a selective yield would not rise far above the value of the unselective MSY. At least in the initial stages of a program age-selective harvesting is seldom worth considering.

Selective harvesting of sexes

When harvesting is selective by age a change in the rate at which one age class is harvested affects the permissible harvesting rates of all age classes. Reciprocal interactions are much weaker between the dynamics of the two sexes (Figure 9.1). In the link between the male segment and female segment of the population there is considerable play that allows us to bend the relationship to our advantage. So long as density is held constant, an increase or decrease in the rates at which males are harvested has no necessary effect on the harvesting rate permissible for females. Most populations contain more males than are needed to fertilize all the females capable of reproduction. Progressive reduction of the proportion of males in the population has little effect on the fecundity of females until a critical threshold is reached. It may be very low, particularly when mating is promiscuous or when a few dominant males monopolize the receptive females. In these circumstances the MSY can be raised well beyond that accruing from unselective harvesting by reducing the proportion of males. Since the production of male recruits per female remains unaffected, the drive towards re-establishment of the stable sex ratio (Section 9.5) must be counteracted by harvesting males at a permanently increased rate.

Two separate factors determine the level to which males can be reduced without impairment of female fecundity. The first is density. Klomp, van Montfort and Tammes (1964) showed that the number of contacts between males and females is proportional to the square root of density and that below a certain threshold the proportion of females fertilized declines at an increasing rate as density is lowered. This effect has little relevance to the design of a harvesting program because the density appropriate to an MSY is always well above the threshold. The second influence on fecundity is the ratio of males to females. A given number of males can fertilize only so many females. When $R = N_m/N_f$ is the ratio of males to females and k is the maximum number of times an average male can copulate during the rutting season, the mean number of copulations per female will be kR. Its value determines the proportion of females in reproductive condition that will be fertilized. A couple of simplified models will make this clear. In the first (model A), females copulate only once during the rutting season and reject all males thereafter. The proportion of females fertilized, P, is $P = kR - g$ where g, an index of untapped male fertility, is the amount by which kR exceeds unity. In model B females are receptive to males whether or not they have copulated previously. The fraction of females that copulate zero, once, twice, and so on times will form a Poisson distribution that determines the proportion of females fertilized as $P = 1 - e^{-kR}$.

Figure 11.4 diagrams the relationships between fecundity and sex ratio implied by the two models. In reality the proportion of females fertilized depends also on such influences as duration of the rutting season, mobility of both sexes and degree of hierarchial dominance, but these complications tend to change only the slope of the curves in Figure 11.4, not their general form.

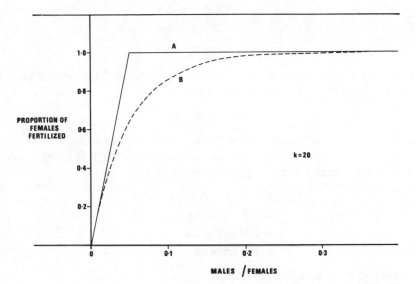

Figure 11.4. The relationship between fecundity and sex ratio when (A) females copulate only once during the rutting season, and (B) females remain receptive whether or not they have copulated previously. In this example an average male is capable of $k = 20$ copulations per season.

The figure indicates that when the population approximates either model A or B there is threshold below which further imbalance of sex ratio adversely affects fecundity and hence the sustained yield. In the diagrammed example the threshold lies at $R = 0.05$ for model A and at about $R = 0.2$ for model B. In practice the threshold must be determined empirically by experimental manipulation of the sex ratio.

We will now investigate the gain of the sex-selective SY over an unselective yield, assuming that the proportion of females, P_f, is held on the safe side of the threshold of breeding impairment.

Suppose a population with an even sex ratio at all ages (i.e. $P_f = 0.5$) is held at a density appropriate to the MSY. It has a fecundity rate of $m = 0.5$ female offspring/female/year, and a mean survival rate per head of $\bar{p} = 0.8$ when unharvested. In this context $\bar{p} = 1 - \bar{n}$, $\bar{n}$ being the mean isolated rate of natural mortality. The population is harvested each year directly after the birth pulse to leave 1000 survivors. Thus it will be reduced by natural mortality to $N\bar{p} = 800$ animals just before the next birth pulse, will be increased by $Npm = 400$ recruits at the birth pulse to a size of 1200, and then an SY of 200 is harvested to return the population to its stable post-harvest size of 1000. Hence

$$\begin{aligned} SY &= N\bar{p} + N\bar{p}m - N \\ &= 800 + 400 - 1000 \\ &= 200 \text{ animals,} \end{aligned}$$

harvested at a rate per head of

$$h = \frac{N\bar{p} + N\bar{p}m - N}{N\bar{p} + N\bar{p}m}$$

$$= 0 \cdot 1667$$

That equation can be simplified by substituting e^{r_p}, the finite rate of increase in the absence of harvesting, for $\bar{p} + \bar{p}m$ to give

$$h = 1 - e^{-r_p},$$

a relationship arrived at earlier in Section 11.2 by another route.

These formulae must now be written specifically for each sex if they are to be used in sex-selective harvesting. The female SY is

$$\begin{aligned} SY_\female &= NP_f\bar{p} + NP_f\bar{p}m - NP_f \\ &= NP_f(\bar{p} + \bar{p}m - 1) \\ &= 500(0 \cdot 8 + 0 \cdot 4 - 1) \\ &= 100 \text{ females.} \end{aligned}$$

They are harvested at a rate per head of

$$\begin{aligned} h_\female &= \frac{\bar{p} + \bar{p}m - 1}{\bar{p} + \bar{p}m} \\ &= \frac{0 \cdot 2}{1 \cdot 2} \\ &= 0 \cdot 1667 \end{aligned}$$

The SY for males is given by

$$\begin{aligned} SY_\male &= N(1 - P_f)\bar{p} + NP_f\bar{p}m - N(1 - P_f) \\ &= N(P_f\bar{p}m - P_f\bar{p} + P_f + \bar{p} - 1) \\ &= 1000(0 \cdot 2 - 0 \cdot 4 + 0 \cdot 5 + 0 \cdot 8 - 1) \\ &= 100 \text{ males} \end{aligned}$$

harvested at a rate per head of

$$\begin{aligned} h_\male &= \frac{N(1 - P_f)\bar{p} + NP_f\bar{p}m - N(1 - P_f)}{N(1 - P_f)\bar{p} + NP_f\bar{p}m} \\ &= 1 - \frac{1 - P_f}{\bar{p}(1 - P_f + P_f m)} \\ &= 1 - \frac{1 - 0 \cdot 5}{0 \cdot 8(1 - 0 \cdot 5 + 0 \cdot 25)} \\ &= 0 \cdot 1667 \end{aligned}$$

Now suppose that the population is held constant as before at a post-harvest size of 1000 animals, but that by sex-selective harvesting the females are held at 600 and the males at 400, i.e. $P_f = 0.6$ rather than 0.5. The female sustained yield is then

$$\begin{aligned} SY_\female &= NP_f(\bar{p} + \bar{p}m - 1) \\ &= (1000 \times 0.6)(0.8 + 0.4 - 1) \\ &= 120 \text{ females,} \end{aligned}$$

and that for males is

$$\begin{aligned} SY_\male &= N(P_f\bar{p}m - P_f\bar{p} + P_f + \bar{p} - 1) \\ &= 1000(0.24 - 0.48 + 0.6 + 0.8 - 1) \\ &= 160 \text{ males.} \end{aligned}$$

The sex-selective rates of harvesting, calculated as before, are $h_\female = 0.1667$ and $h_\male = 0.285$. Note that only the rate for males has been changed by this treatment. A sustained yield is maintained through sex-selective harvesting by holding constant the harvesting rate (but not the yield) for females while increasing the rate at which males are harvested.

The SY from a population of fixed size increases linearly as P_f rises:

$$SY = (2N\bar{p}m)P_f - N(1 - \bar{p}),$$

where $-N(1 - \bar{p})$ is the y-intercept and $2N\bar{p}m$ is the slope. The x-intercept, $(1 - \bar{p})/2\bar{p}m$, gives the value of P_f below which no degree of sex-selective harvesting can hold both P_f and population size constant.

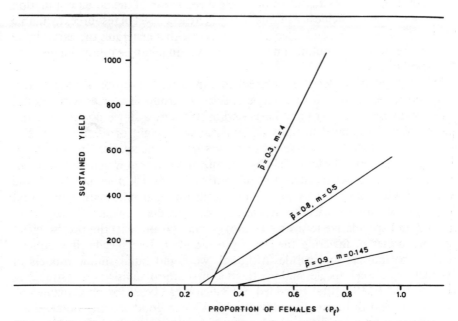

Figure 11.5. The relationship between sustained yield and the sex ratio to which the population is held by sex-selective harvesting.

Figure 11.5 shows the order of gain that might be expected when a game bird (parameters of $\bar{p} = 0.3$, $m = 4$), an ungulate ($\bar{p} = 0.8$, $m = 0.5$), and a large mammals such as an elephant ($\bar{p} = 0.9$, $m = 0.145$), is harvested to raise the standing P_f beyond 0.5. In each case the rise in yield is considerable. This model is simplistic as well as simple. Its greatest abstraction is the implicit assumption that $\bar{p}$ is independent of age distribution. This is not relevant to the females whose age distribution is not changed by sex-selective harvesting, but it cannot be ignored for the males whose mean age in the standing population decreases as the sex-ratio is shifted in favour of females. Such a shift is likely to load the males into those young age classes subjected to a higher than average rate of natural mortality. Hence $\bar{p}$ is likely to decline as P_f is increased and the regressions drawn as straight lines in Figure 11.5 are more likely to curve convexly. But within the practical limits imposed on strong sex-selective harvesting under field conditions, the formulae given here will usually be close enough for estimating sex-selective yields.

Selective harvesting of species

When a harvested species shares resources with several other species, the relationship between numbers in one year and numbers in the next is unlikely to be as simple as that diagrammed in Figure 11.1. A harvest-induced reduction of the target population's density may cause an increase in the density of one or more unharvested species inhabiting the same area if diets or habitat requirements overlap. In this situation a permanent reduction of the target population does not necessarily achieve a permanently increased level of resources available to each of its members; and since it is this increase that generates the harvestable surplus the previously outlined models of harvesting may be totally inadequate for estimating MSY.

When population densities interact in this way, harvesting should not, if at all possible, be limited to a single species. A group of species with similar ecological requirements can be harvested as if it were a single population, and the MSY is determined in units of total carcase weight or cash return. The exercise is analogous to sex-selective harvesting of one species. In both cases we seek to maximize the total yield by optimizing the rate at which each segment of the population is harvested. But in contrast to sex-selective harvesting, the effects of which are broadly predictable from minimal information on social organization, species-selective harvesting can produce surprises. Except in general and speculative terms we cannot predict (at present) the precise effect on one species of lowering the density of another. Interspecific interactions tend to be indirect and subtle. Although we could build simple models of interaction between species and use them as first approximations for estimating the combined MSY (that provided by Riffenburg (1969) for maximizing the combined yield of herring, anchovy and hake is a good starting point) these are not substitutes for determining interactive effects directly by experimental manipulation.

11.2.6 Harvesting in a fluctuating environment

The methods previously outlined for estimating MSY can cope with only a moderate degree of year-to-year fluctuation in a population's conditions of life. They will be adequate for most populations. Some populations, however, have a 'boom and bust' economy, increasing when conditions are favourable and crashing to low levels when the environment deteriorates. In this category can be placed most of the species living in deserts or semi-deserts, many species of oceanic fish, and several vertebrates of the northern tundra.

The simplifying assumption of a steady density, which at best is a dangerous abstraction, almost completely parts company with reality when applied to a 'boom and bust' population. Similarly, the use of density as an index of the availability of resources per head, and by extension as an index of r_p, cannot be justified when a population's size fluctuates wildly. In these circumstances density and resources are seldom balanced and one cannot be predicted with confidence from the other.

The problem of harvesting populations in a fluctuating environment can be approached from two directions. The variability could simply be ignored and the MSY estimated as if the population were stable in size, the 'steady density' being taken as the population's size averaged over several years. This is called a 'mean strategy'. Alternatively, a 'tracking strategy' can be used. The rate of harvesting is increased as r rises and harvesting is curtailed or suspended when r is negative. In this way the harvesting strategy tracks the population, changing as the population changes.

A mean strategy has many advantages, particularly when the amplitude of fluctuation in numbers is not too great and the average time between fluctuations is not too long. Economically, it is preferable to a tracking strategy because it minimizes market fluctuations. For this reason it is favoured by fisheries biologists, who have shown that a mean strategy is reasonably successful even when recruitment fluctuates markedly and unsystematically from year to year. Its disadvantages lie in the danger of cropping too heavily at low density to the extent that the population's ability to increase is seriously impaired at the time conditions again become favourable. The design of a mean strategy requires considerable information. It is an unwise choice when little is known about the magnitude of environmental fluctuations or about the population's dynamics. Nor is it a strategy for the faint-hearted; there is always a strong temptation to shift from a mean to a tracking strategy when numbers are low.

A tracking strategy is safer, particularly when little is known about the population and its environment, but it results in fluctuating market prices and labour requirements, and generally a lower average yield. However, it is the only workable strategy when environmental fluctuations are substantial and when their mean periodicity exceeds about five years.

When a harvesting program is in its initial stages a mean strategy should not be used unless it is based on considerable information. A tracking strategy is appropriate until enough data have accumulated to allow a decision between the two approaches.

11.2.7 Discussion of harvesting

In the simplest case the maximization of yield is achieved by these steps:

1. Estimate steady size K.
2. Harvest a constant number of animals in each of several years and estimate r_m from the rate at which the population declines.
3. Estimate the population size appropriate to MSY harvesting (from K) and the instantaneous rate of harvesting, H, appropriate to a population of this size (from r_m).
4. Harvest the population down to the MSY density and then continue harvesting at rate H if the population is harvested throughout the year or at rate h if the season of harvesting is restricted.
5. If the sexes can be differentiated in the field, determine the sex ratio below which further imbalance affects fecundity.
6. Harvest males at a higher rate to fix the sex ratio at optimal disparity.

When outlined in this way, MSY harvesting appears as a straightforward exercise in field ecology and applied arithmetic, demanding no more than adequate equipment and a modicum of technical competence. The impression is false. Analytical techniques used to calculate MSY provide approximations, not exact solutions. At each step of the treatment the population's reaction must be monitored to determine empirically whether the calculated MSY will stabilize the population at the calculated MSY density. Most methods by which the MSY is estimated are based on simplified models of a population's reaction to lowered density. Although we can be quite certain that the population will not react precisely as the model predicts, we cannot foretell whether the model will be close enough for practical purposes or whether it will be inappropriate.

The logistic model, the one featured in previous sections, will not be appropriate in all cases. Implicit in its use is the assumption that the rate of production of food or of any other consumable resource, is independent of the number of animals using it; the animals consume a proportion of the interest but leave the capital intact. In fact, most populations of grazers and browsers both influence and are influenced by the number of species, the kinds of species and the density of each species of the plants growing in their area. An artificial reduction in density of animals will, if maintained for long enough, induce changes in plant composition. Under a regime of MSY harvesting the density of the population will, after a few years, commence to drift upward in response to an increase in capital food supply. The effect is clearly demonstrated by the trend in annual tallies of large mammals shot in Botswana in an attempt to control tsetse fly (Child, Smith and von Richter 1970). Of the several ways of dealing with this effect the simplest is to continue harvesting at the rate $H = r_m/2$ per year, thereby taking the same percentage, but a higher absolute yield each year, as the population increases. Density will eventually stabilize at a higher level, being about half the true steady density, K, the initial estimate of which is usually too low.

A major cause of inaccuracy in estimating the MSY is the disruption in a population's social organization caused by the harvesting. Groups of animals are seldom random aggregations and they often react in unexpected ways to the loss of some of their members. A social species should not be harvested until its social organization is fully understood. We must be able to predict both whether harvesting will disrupt the organization of social groups and the effect of this disruption on r_p.

Throughout the previous sections the MSY, whether it be computed as numbers, value, or carcase weight, has been treated as if it were the only rational goal of a harvesting program. Sometimes it is; often it is not. The optimal density of the harvested population may be above that appropriate to an MSY when the population is managed for those who wish to look at animals as well as for those that wish to hunt them. Alternatively, the optimum density may be below the MSY density if the population is managed to produce large trophy animals or if the animals are interfering with an attempt to increase ground cover for flood and erosion control.

Occasionally the best management may be no management at all. Management activities can sometimes conflict with the intangible values that a population provides, to the extent that they are counter-productive. In the community that raised me, a red deer with 40 inches of antler was a greater prize than a ton of meat; and a stag whose antlers measured 45 inches was a trophy beyond price. A wildlife manager sufficiently skilled and sufficiently blind could descend on this community—it still exists—and shape that herd of deer to produce large stags with notable frequency. He would not be thanked. In that ethos a great stag is born of the mountains and the rain, it is not a manufactured thing. It must be worthy of its own death. In working his art to give that community what he thought it wanted, in plugging constants into a FORTRAN program to calculate the parameters of an optimal equilibrium that existed solely in his own mind, the wildlife manager would be negating a system of values that hardly impinges at all upon his own. We must be cautious of values that summarize neatly as numbers. They may be insubstantial.

Even when an MSY appears a valid goal there are situations in which it is economically impracticable. The expenses of harvesting rises steeply as density is reduced and the greatest net revenue may often come from a density higher than that appropriate to the MSY. In particular, sex-selective harvesting calls for a higher outlay in skill and money than does unselective cropping. The gain in off-take must be measured carefully against the reduction in net revenue per head of yield.

These cautionary remarks emphasize that MSY harvesting is fraught with unexpected difficulties. The calculation of an MSY must be treated as a first approximation and the effect of the harvesting must be followed carefully to allow fine adjustment towards the optimum. Beverton and Holt (1957: 436) were discussing fish in the following passage but their remarks are equally applicable to the harvesting of birds and mammals: 'it is the changes produced in the fisheries by the regulations themselves, whether they be the first or the last in the series leading towards the optimum, that provide the opportunity of obtaining, by research, just the information that may have been lacking previously. Thus the approach towards optimum fishing, and the increase in knowledge of where the optimum lies, can be two simultaneous and complementary advances; the benefits to the fisheries of such progress can hardly be exaggerated'.

11.3 CONTROL

Most populations extinguished by man were eliminated by accident, often in spite of vigorous efforts to avert the extinction. In contrast, most premeditated attempts to destory a population have been unsuccessful. The paradox is less puzzling when it is realized that unplanned exterminations are usually caused by a change in the animals' habitat whereas the planned attempts are usually aimed directly at the animals themselves. The message is clear: populations are more vulnerable to a manipulation of their habitat than they are to a direct manipulation of their numbers. An environmental change tends to affect the quality of one or more environmental components and when this change is deleterious the population cannot necessarily adjust to it by lowering density as it can when the change affects only the quantity of a resource. A population attacked frontally by shooting or poisoning does not have to contend with a deteriorating habitat. Quite the reverse. The reduction in density occasioned by the control measures leaves the quality of resources intact while increasing the quantity available to each surviving animal. The control campaign automatically boosts the value of r_p. This is precisely the aim of a harvesting program. Most control operations are, in reality, informal harvesting programs that take a sustained yield.

One of the most elaborate control operations mounted against vertebrates was aimed at eliminating large mammals from several African tsetse fly areas. Child, Smith and von Richter (1970) summarized the result on the Mauna Front in northern Botswana: 'With the possible exception of roan antelope, which were common in the area at the start of the control operation and which have declined in numbers in much of their former range in northern Botswana, no species was eliminated by 23 years of hunting. Only tsessebe, wildebeest and kudu were shot in reduced numbers in later years and there is evidence that the declining trends in annual kills of the last two was due to factors other than hunting. All other species showed an upward trend, which, supported by other types of evidence, could not be reconciled with diminishing populations in the effective hunting area.'

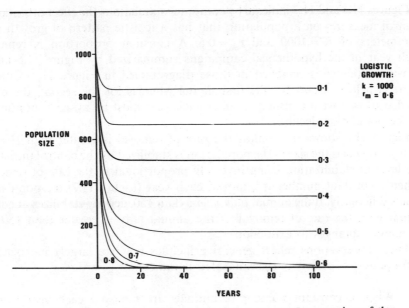

Figure 11.6. Trend of population size when a constant proportion of the population is removed each year. The maximum sustained yield is removed at a rate of 0·3.

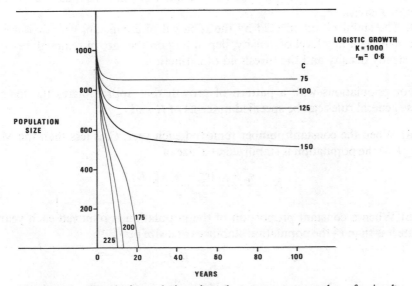

Figure 11.7. Trend of population size when a constant number of animals are removed each year. The maximum sustained yield is 150 per year.

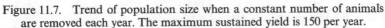

Figures 11.6 and 11.7 indicate by computer simulations the effect of sustained control measures on a population that has a logistic pattern of growth with parameters of $K = 1000$ and $r_m = 0{\cdot}6$. A constant proportion is removed each year in the hypothetical campaigns summarized by Figure 11.6 and a constant number is removed in those diagrammed in Figure 11.7. Control effort is held constant for the first model whereas for the second the effort needed to remove a constant number each year must be raised continuously as the population declines.

Figure 11.6 shows that unless the rate of removal is greater than $H = 0{\cdot}6$ on a yearly basis the size of the population is stabilized by the control measures, the level of stabilization being inversely proportional to the rate of removal. When a constant number is removed each year (Figure 11.7) the population reacts differently. If the annual tally is less than 150, density stabilizes at equilibrium with the rate of removal. If the annual tally is greater than 150 the population plunges to extinction.

These observations reflect general principles that are largely independent of the pattern of growth or the value of its parameters:

1. When a constant number of animals are removed each year from a population initially at steady density, the population will be stabilized by the control operations unless the annual tally exceeds the MSY.
2. The level at which density is stabilized by the removal of a constant number each year is always equal to or above the density from which the MSY is harvested.
3. Density is stabilized by the removal each year of a constant proportion of the population providing that proportion does not exceeds the intrinsic rate of increase, r_m.
4. The stabilization effected by the removal of a constant proportion each year can be at any level of density, depending on the rate of removal, between the steady density and the threshold of extinction.

For populations with a pattern of growth that approximates the logistic, these general rules can be specified in terms of K and r_m:

(a) When the constant number removed each year, C, is less than the MSY of $r_m K/4$ the population is stabilized at a size of

$$N = \frac{r_m + \sqrt{(r_m^2 - 4Cr_m/K)}}{2r_m/K}.$$

(b) When a constant proportion of the population is removed each year at a rate less than r_m the population stabilizes at a size of

$$N = K - \frac{KH}{r_m}$$

where H is the instantaneous rate of removal.

(c) If animals are removed at a constant rate greater than r_m, or if a constant number greater than $r_m K/4$ is removed each year, the population will eventually become extinct.

In the sense used here 'stabilization' implies no more than a temporary equilibrium. If, over the long run, the capital level of resources is itself in balance with animal density, a lowering of this density will stimulate a rise in the absolute level of these resources. The population will then increase in size if the effort expended on control is held constant. The report by Child, Smith and von Richter (1970) quoted previously in this section described a general rise in the yield per unit effort after many years of control measures aimed at exterminating populations of large mammals. Inescapably, the rise in yield reflects an increased density of animals. That the trend is a reaction to a rising level of capital resources (particularly food) is less certain, but only marginally so.

As emphasized previously, most attempts to eliminate populations have ended up as unplanned and undesired exercises in sustained-yield harvesting. On balance this is a good thing because, with the benefit of hindsight, it can be shown that most of these attempts were ecologically unjustified and economically impracticable.

I must admit here to a personal bias. Although I have no objection to control measures aimed at holding the density of a population below K, provided there is good reason for doing so, I react to the suggestion that a population should be exterminated. On this I am flexible—I have supported the extermination of feral goats on certain small oceanic islands, for example—but in general I have neither the necessary knowledge nor prescience, and hence the right, to make or subserve decisions of this magnitude. When, as in this case, the penalty for being wrong vastly outweighs the reward flowing from a correct decision, irreversible actions should not be contemplated lightly. I will outline here the economic arguments in favour of this position even though I reached it by an entirely different path and hold it for reasons that have nothing to do with economics. I do not expect all biologists to share my personal view but I do expect those that oppose it to consider the economic implications of an attempt at extermination. The task is expensive and difficult. To launch an operation of this kind without first costing out the benefit and the expense is not only irresponsible administratively and technically, it is economically stupid. Minimum data for costing are estimates of r_m and K, a knowledge of the expense incurred by one unit of effort and the proportionate reduction in density obtained thereby.

The computer population of Figure 11.6 will serve as an example. One unit of effort will be defined as that needed to remove animals at the annual instantaneous rate of $H = 0.1$. For the kind of situation I have in mind, the elimination of a small population of medium-sized ungulates, this rate would be achieved by one hunter working for one year. A unit of effort can therefore be expressed as one man/year of hunting. The rate of removal must obviously exceed $r_m = 0.6$ per year or the operation will simply extract a sustained yield.

At $H = 0.65$ an average of 6·5 men must be employed each year to provide the required hunting pressure. These men will take 72 years and the expenditure of 468 man/years of effort to eliminate the population. At $H = 0.7$ the population is reduced to the threshold of extinction after 42 years and 294 man/years of effort. The cost of the operation declines as the rate of removal is increased: at $H = 0.8$ only 184 man/years are required. These rates of removal are much higher than those normally considered. Only a few of the attempts at extermination that I have investigated utilized a rate of removal above about $H = 0.1$ per year, a rate that will seldom achieve more than an inefficient sustained yield.

The estimate of the considerable effort required to exterminate the small population simulated by the computer model is impressive enough in itself. However, it is actually a gross underestimate because the model lacks two important attributes of a real population. Firstly, unless the population is eliminated swiftly, the rise in the capital level of resources triggered by the decline in density will bring about a rise in r_p and a consequent increase in the effort needed to reduce the population further. Secondly, the model requires that one unit of hunting effort removes a constant proportion of the population irrespective of the population's density. In the example it removes 100 animals when $N = 1000$, 50 when $N = 500$ and 10 when $N = 100$. In practice this assumption is probably reasonable down to a population size of $K/2$ but below that it rapidly comes unstuck. Animals are not equally catchable and a control operation progressively culls out those least able to adapt to it. The survivors at low density comprise a large proportion of animals that are either extremely wary or that have home ranges in areas that are difficult to hunt across. Hence a unit of effort expended when density is low will reduce the population by a lesser fraction than when density is high.

Control is usually a more practicable procedure than is extermination and it lends itself to more rational planning. Costing is again important. The benefit of holding a population at a given density must be weighed against the expense, the required rate of removal being calculated from estimates of r_m and K and the cost being calculated from the estimate of effort needed to maintain this rate. I have no knowledge of any control program being preceded by such a study. At present (1975) several countries are financing control schemes that appear to an outside observer to consume more money than could conceivably return as benefits. In this I might well be wrong, but since cost-benefit analyses are available neither to the governments concerned, nor to me, my curiosity remains unsatisfied. Since control of a population by hunting, trapping or poisoning is an exercise in sustained-yield harvesting, whether or not the yield is utilized, the analyses appropriate to its planning are precisely those used in a harvesting program. Direct control and harvesting are operationally identical. The efficiency of the first is determined by the prevailing level of technical competence in the techniques of the second.

Control by way of pathogens is a seductive notion and the success that entomologists can claim for it increases its attractiveness. Unfortunately there is but one example of the successful control of a vertebrate population by pathogens—the use of myxomatosis against rabbits. This success story (at least it can be rated a success in Australia) is described in detail by Fenner and Ratcliffe (1965). The virus was first observed in 1896 as a lethal infection of laboratory rabbits (*Oryctolagus*) in South America. It exists naturally as a non-lethal pathogen of American rabbits *Sylvilagus*. The European rabbit *Oryctolagus* evolved in an environment that had never contained viruses of the myxomatosis group. It proved highly susceptible on initial contact with the virus. As Burnet (1962) emphasized, the balance between a freely spreading virus and a host for which it is highly lethal cannot be maintained for long since the reduction in host numbers will prevent the effective spread of the disease and thereby eliminate the virus. Predictably, within two or three years after introduction of the virus into Australia there was a significant decrease in virulence, and within eight years genetic resistance made its appearance amongst the rabbits.

This example underlines the difficulty of using pathogens to control vertebrates. Firstly we must find a pathogen that has never been in contact with the target species but which is hosted by a closely related species. Secondly we must be certain that the pathogen will attack only the target species and not spread to others inhabiting the same area. And thirdly we must accept that the pathogen will provide only temporary respite because both it and the target species will, through the agency of massive natural selection, quickly reach an accommodation. These difficulties count heavily against the search for a pathogen to control a vertebrate population. The chance of finding a suitable disease is very low, and even when one is found it will not provide a permanent solution to the problem.

Whenever possible, attempts at lowering density should be aimed towards manipulating the habitat. This method has few of the drawbacks of direct control or control by pathogens. Most populations are susceptible to changes in a key component of the habitat and the lowered density effected thereby can usually be fixed with little effort. The European rabbit can again serve as an example. In New Zealand the rabbit could not be controlled by myxomatosis, as it was in Australia, because mosquito vectors were not sufficiently numerous (Filmer 1953). Dense concentrations of rabbits were usually attacked by laying 1080 poison from the air. Howard (1959) showed that when density had been lowered by these methods and the reduced grazing by rabbits had allowed the grass to reform a continuous sward, the rabbits were incapable of again erupting. Thereafter rabbits could be controlled most effectively by regulating sheep grazing and by applying fertilizer to ensure that the sward remained in healthy condition and was not grazed too short. The rabbit being essentially a desert animal, it cannot maintain high densities except in desert conditions or where these conditions are created artificially.

The treatment of a population by changing to its detriment the key components of its habitat is the most powerful and elegant technique of population control.

Appendix

x	0	1	2	3	4	5	6	7	8	9
.0	1.0000	1.0101	1.0202	1.0305	1.0408	1.0513	1.0618	1.0725	1.0833	1.0942
.1	1.1052	1.1163	1.1275	1.1388	1.1503	1.1618	1.1735	1.1853	1.1972	1.2092
.2	1.2214	1.2337	1.2461	1.2586	1.2712	1.2840	1.2969	1.3100	1.3231	1.3364
.3	1.3499	1.3634	1.3771	1.3910	1.4049	1.4191	1.4333	1.4477	1.4623	1.4770
.4	1.4918	1.5068	1.5220	1.5373	1.5527	1.5683	1.5841	1.6000	1.6161	1.6323
.5	1.6487	1.6653	1.6820	1.6989	1.7160	1.7333	1.7507	1.7683	1.7860	1.8040
.6	1.8221	1.8404	1.8589	1.8776	1.8965	1.9155	1.9348	1.9542	1.9739	1.9937
.7	2.0138	2.0340	2.0544	2.0751	2.0959	2.1170	2.1383	2.1598	2.1815	2.2034
.8	2.2255	2.2479	2.2705	2.2933	2.3164	2.3396	2.3632	2.3869	2.4109	2.4351
.9	2.4596	2.4843	2.5093	2.5345	2.5600	2.5857	2.6117	2.6379	2.6645	2.6912
1.0	2.7183	2.7456	2.7732	2.8011	2.8292	2.8577	2.8864	2.9154	2.9447	2.9743
1.1	3.0042	3.0344	3.0649	3.0957	3.1268	3.1582	3.1899	3.2220	3.2544	3.2871
1.2	3.3201	3.3535	3.3872	3.4212	3.4556	3.4903	3.5254	3.5609	3.5966	3.6328
1.3	3.6693	3.7062	3.7434	3.7810	3.8190	3.8574	3.8962	3.9354	3.9749	0149
1.4	4.0552	4.0960	4.1371	4.1787	4.2207	4.2631	4.3060	4.3492	4.3929	4.4371
1.5	4.4817	4.5267	4.5722	4.6182	4.6646	4.7115	4.7588	4.8066	4.8550	4.9037
1.6	4.9530	5.0028	5.0531	5.1039	5.1552	5.2070	5.2593	5.3122	5.3656	5.4195
1.7	5.4739	5.5290	5.5845	5.6407	5.6973	5.7546	5.8124	5.8709	5.9299	5.9895
1.8	6.0496	6.1104	6.1719	6.2339	6.2965	6.3598	6.4237	6.4883	6.5535	6.6194
1.9	6.6859	6.7531	6.8210	6.8895	6.9588	7.0287	7.0993	7.1707	7.2427	7.3155

Appendix 1A (*Contd.*)

X	0	1	2	3	4	5	6	7	8	9
2.0	7.3891	7.4633	7.5383	7.6141	7.6906	7.7679	7.8460	7.9248	8.0045	8.0849
2.1	8.1662	8.2482	8.3311	8.4149	8.4994	8.5849	8.6711	8.7583	8.8463	8.9352
2.2	9.0250	9.1157	9.2073	9.2999	9.3933	9.4877	9.5831	9.6794	9.7767	9.8749
2.3	9.9742	10.0744	10.1757	10.2779	10.3812	10.4856	10.5910	10.6974	10.8049	10.9135
2.4	11.0232	11.1340	11.2459	11.3589	11.4730	11.5883	11.7048	11.8224	11.9413	12.0613
2.5	12.1825	12.3049	12.4286	12.5535	12.6797	12.8071	12.9358	13.0658	13.1971	13.3298
2.6	13.4637	13.5991	13.7357	13.8738	14.0132	14.1540	14.2963	14.4400	14.5851	14.7317
2.7	14.8797	15.0293	15.1803	15.3329	15.4870	15.6426	15.7998	15.9586	16.1190	16.2810
2.8	16.4446	16.6099	16.7769	16.9455	17.1158	17.2878	17.4615	17.6370	17.8143	17.9933
2.9	18.1741	18.3568	18.5413	18.7276	18.9158	19.1060	19.2980	19.4919	19.6878	19.8857
3.0	20.0855	20.2874	20.4913	20.6972	20.9052	21.1153	21.3276	21.5419	21.7584	21.9771
3.1	22.1980	22.4210	22.6464	22.8740	23.1039	23.3361	23.5706	23.8075	24.0468	24.2884
3.2	24.5325	24.7791	25.0281	25.2797	25.5337	25.7903	26.0495	26.3113	26.5758	26.8429
3.3	27.1126	27.3851	27.6604	27.9383	28.2191	28.5027	28.7892	29.0785	29.3708	29.6660
3.4	29.9641	30.2652	30.5694	30.8766	31.1870	31.5004	31.8170	32.1367	32.4597	32.7859
3.5	33.1155	33.4483	33.7844	34.1240	34.4669	34.8133	35.1632	35.5166	35.8735	36.2341
3.6	36.5982	36.9661	37.3376	37.7128	38.0918	38.4747	38.8613	39.2519	39.6464	40.0448
3.7	40.4473	40.8538	41.2644	41.6791	42.0980	42.5211	42.9484	43.3801	43.8160	44.2564
3.8	44.7012	45.1504	45.6042	46.0625	46.5255	46.9931	47.4654	47.9424	48.4242	48.9109
3.9	49.4024	49.8990	50.4004	50.9070	51.4186	51.9354	52.4573	52.9845	53.5170	54.0549

4.0	54.5982	55.1469	55.7011	56.2609	56.8263	57.3975	57.9743	58.5570	59.1455	59.7399
4.1	60.3403	60.9467	61.5592	62.1779	62.8028	63.4340	64.0715	64.7155	65.3659	66.0228
4.2	66.6863	67.3565	68.0335	68.7172	69.4079	70.1054	70.8100	71.5216	72.2404	72.9665
4.3	73.6998	74.4405	75.1886	75.9443	76.7075	77.4785	78.2571	79.0436	79.8380	80.6404
4.4	81.4509	82.2695	83.0963	83.9314	84.7749	85.6269	86.4875	87.3567	88.2347	89.1214
4.5	90.0171	90.9218	91.8356	92.7586	93.6908	94.6324	95.5835	96.5441	97.5144	98.4944
4.6	99.4843	100.4841	101.4940	102.5141	103.5443	104.5850	105.6361	106.6977	107.7701	108.8532
4.7	109.9472	111.0522	112.1683	113.2956	114.4342	115.5843	116.7459	117.9192	119.1044	120.3014
4.8	121.5104	122.7316	123.9651	125.2110	126.4694	127.7404	129.0242	130.3209	131.6307	132.9536
4.9	134.2898	135.6394	137.0026	138.3795	139.7702	141.1750	142.5938	144.0269	145.4744	146.9364
5.0	148.4132	149.9047	151.4113	152.9330	154.4700	156.0225	157.5905	159.1743	160.7741	162.3899
5.1	164.0219	165.6704	167.3354	169.0171	170.7158	172.4315	174.1645	175.9148	177.6828	179.4686
5.2	181.2722	183.0941	184.9342	186.7928	188.6701	190.5663	192.4815	194.4160	196.3699	198.3434
5.3	200.3368	202.3502	204.3839	206.4380	208.5127	210.6083	212.7249	214.8629	217.0223	219.2034
5.4	221.4064	223.6316	225.8791	228.1492	230.4422	232.7582	235.0974	237.4602	239.8467	242.2572
5.5	244.6919	247.1511	249.6350	252.1439	254.6780	257.2376	259.8228	262.4341	265.0716	267.7356
5.6	270.4264	273.1442	275.8894	278.6621	281.4627	284.2915	287.1486	290.0345	292.9494	295.8936
5.7	298.8674	301.8711	304.9049	307.9693	311.0644	314.1907	317.3483	320.5377	323.7592	327.0130
5.8	330.2996	333.6191	336.9721	340.3587	343.7793	347.2344	350.7241	354.2490	357.8092	361.4053
5.9	365.0375	368.7062	372.4117	376.1545	379.9349	383.7533	387.6101	391.5057	395.4404	399.4146

Appendix 1B. Exponential functions of $-x$

x	0	1	2	3	4	5	6	7	8	9
-.0	1.0000	.9900	.9802	.9704	.9608	.9512	.9418	.9324	.9231	.9139
.1	.9048	.8958	.8869	.8781	.8694	.8607	.8521	.8437	.8353	.8270
.2	.8187	.8106	.8025	.7945	.7866	.7788	.7711	.7634	.7558	.7483
.3	.7408	.7334	.7261	.7189	.7118	.7047	.6977	.6907	.6839	.6771
.4	.6703	.6637	.6570	.6505	.6440	.6376	.6313	.6250	.6188	.6126
.5	.6065	.6005	.5945	.5886	.5827	.5769	.5712	.5655	.5599	.5543
.6	.5488	.5434	.5379	.5326	.5273	.5220	.5169	.5117	.5066	.5016
.7	.4966	.4916	.4868	.4819	.4771	.4724	.4677	.4630	.4584	.4538
.8	.4493	.4449	.4404	.4360	.4317	.4274	.4232	.4190	.4148	.4107
.9	.4066	.4025	.3985	.3946	.3906	.3867	.3829	.3791	.3753	.3716
1.0	.3679	.3642	.3606	.3570	.3535	.3499	.3465	.3430	.3396	.3362
1.1	.3329	.3296	.3263	.3230	.3198	.3166	.3135	.3104	.3073	.3042
1.2	.3012	.2982	.2952	.2923	.2894	.2865	.2837	.2808	.2780	.2753
1.3	.2725	.2698	.2671	.2645	.2618	.2592	.2567	.2541	.2516	.2491
1.4	.2466	.2441	.2417	.2393	.2369	.2346	.2322	.2299	.2276	.2254
1.5	.2231	.2209	.2187	.2165	.2144	.2122	.2101	.2080	.2060	.2039
1.6	.2019	.1999	.1979	.1959	.1940	.1920	.1901	.1882	.1864	.1845
1.7	.1827	.1809	.1791	.1773	.1755	.1738	.1720	.1703	.1686	.1670
1.8	.1653	.1637	.1620	.1604	.1588	.1572	.1557	.1541	.1526	.1511
1.9	.1496	.1481	.1466	.1451	.1437					

	.00	.01	.02	.03	.04	.05	.06	.07	.08	.09
2.0	.1353	.1340	.1327	.1313	.1300	.1287	.1275	.1262	.1249	.1237
2.1	.1225	.1212	.1200	.1188	.1177	.1165	.1153	.1142	.1130	.1119
2.2	.1108	.1097	.1086	.1075	.1065	.1054	.1044	.1033	.1023	.1013
2.3	.1003	.0993	.0983	.0973	.0963	.0954	.0944	.0935	.0926	.0916
2.4	.0907	.0898	.0889	.0880	.0872	.0863	.0854	.0846	.0837	.0829
2.5	.0821	.0813	.0805	.0797	.0789	.0781	.0773	.0765	.0758	.0750
2.6	.0743	.0735	.0728	.0721	.0714	.0707	.0699	.0693	.0686	.0679
2.7	.0672	.0665	.0659	.0652	.0646	.0639	.0633	.0627	.0620	.0614
2.8	.0608	.0602	.0596	.0590	.0584	.0578	.0573	.0567	.0561	.0556
2.9	.0550	.0545	.0539	.0534	.0529	.0523	.0518	.0513	.0508	.0503
3.0	.0498	.0493	.0488	.0483	.0478	.0474	.0469	.0464	.0460	.0455
3.1	.0450	.0446	.0442	.0437	.0433	.0429	.0424	.0420	.0416	.0412
3.2	.0408	.0404	.0400	.0396	.0392	.0388	.0384	.0380	.0376	.0373
3.3	.0369	.0365	.0362	.0358	.0354	.0351	.0347	.0344	.0340	.0337
3.4	.0334	.0330	.0327	.0324	.0321	.0317	.0314	.0311	.0308	.0305
3.5	.0302	.0299	.0296	.0293	.0290	.0287	.0284	.0282	.0279	.0276
3.6	.0273	.0271	.0268	.0265	.0263	.0260	.0257	.0255	.0252	.0250
3.7	.0247	.0245	.0242	.0240	.0238	.0235	.0233	.0231	.0228	.0226
3.8	.0224	.0221	.0219	.0217	.0215	.0213	.0211	.0209	.0207	.0204
3.9	.0202	.0200	.0198	.0196	.0194	.0193	.0191	.0189	.0187	.0185

Appendix 1B (Contd.)

X	0	1	2	3	4	5	6	7	8	9
4.0	.0183	.0181	.0180	.0178	.0176	.0174	.0172	.0171	.0169	.0167
4.1	.0166	.0164	.0162	.0161	.0159	.0158	.0156	.0155	.0153	.0151
4.2	.0150	.0148	.0147	.0146	.0144	.0143	.0141	.0140	.0138	.0137
4.3	.0136	.0134	.0133	.0132	.0130	.0129	.0128	.0127	.0125	.0124
4.4	.0123	.0122	.0120	.0119	.0118	.0117	.0116	.0114	.0113	.0112
4.5	.0111	.0110	.0109	.0108	.0107	.0106	.0105	.0104	.0103	.0102
4.6	.0101	.0100	.0099	.0098	.0097	.0096	.0095	.0094	.0093	.0092
4.7	.0091	.0090	.0089	.0088	.0087	.0087	.0086	.0085	.0084	.0083
4.8	.0082	.0081	.0081	.0080	.0079	.0078	.0078	.0077	.0076	.0075
4.9	.0074	.0074	.0073	.0072	.0072	.0071	.0070	.0069	.0069	.0068
5.0	.0067	.0067	.0066	.0065	.0065	.0064	.0063	.0063	.0062	.0062
5.1	.0061	.0060	.0060	.0059	.0059	.0058	.0057	.0057	.0056	.0056
5.2	.0055	.0055	.0054	.0054	.0053	.0052	.0052	.0051	.0051	.0050
5.3	.0050	.0049	.0049	.0048	.0048	.0047	.0047	.0047	.0046	.0046
5.4	.0045	.0045	.0044	.0044	.0043	.0043	.0043	.0042	.0042	.0041
5.5	.0041	.0040	.0040	.0040	.0039	.0039	.0038	.0038	.0038	.0037
5.6	.0037	.0037	.0036	.0036	.0036	.0035	.0035	.0034	.0034	.0034
5.7	.0033	.0033	.0033	.0032	.0032	.0032	.0032	.0031	.0031	.0031
5.8	.0030	.0030	.0030	.0029	.0029	.0029	.0029	.0028	.0028	.0028
5.9	.0027	.0027	.0027	.0027	.0026	.0026	.0026	.0026	.0025	.0025

Appendix 2. FORTRAN program: computation of rate of increase, r, from a schedule of age-specific $l_x m_x$ values.

```
C
C       ESTIMATING EXPONENTIAL RATE OF INCREASE FROM THE LIFE TABLE AND
C       THE FECUNDITY TABLE.  DATA ARE ARRAY V OF SURVIVAL X FECUNDITY
C       PRODUCTS FOR EACH AGE BEGINNING AT ZERO AGE. LAST DATA CARD HAS
C       99. BEGINNING FROM COLUMN 1.
C
        DIMENSION V(16)
   1    READ (5, 2) V
   2    FORMAT (16F5.3)
        IF (V(1) .EQ. 99.) GO TO 8
        SUM1 = 0.
        SUM2 = 0.
        DO 3 I = 1, 16
        SUM1 = SUM1 + V(I)
   3    SUM2 = SUM2 + V(I) * (I-1)
        R = ALOG(SUM1) * SUM1 / SUM2
   4    R = R + .00001
        SUM = 0.
        DO 5 I = 1, 16
   5    SUM = SUM + V(I) / EXP(R) ** (I-1)
        IF (SUM - 1.) 6, 6, 4
   6    WRITE (6, 7) R, SUM
   7    FORMAT (/// 20X, 4HR = , F6.4/ 20X, 12HSUMMATION = , F6.4//)
        GO TO 1
   8    CONTINUE
        STOP
        END
```

N.B. For most jobs the 0.00001 in statement 4 can be reduced to 0.001 without loss of accuracy, thereby cutting machine time by a factor of 100.

Appendix 3.   FORTRAN program: computation of population size from frequencies of capture.

```
C
C         ESTIMATING POPULATION SIZE FROM CAPTURE FREQUENCIES.   THREE
C         DIFFERENT MODELS ARE USED - NEGATIVE BINOMIAL, GEOMETRIC AND
C         POISSON DISTRIBUTIONS, EACH IN TRUNCATED FORM. DATA ARE ARRAY
C         FREQ OF CAPTURE FREQUENCIES, I.E. NUMBER OF ANIMALS CAPTURED
C         ONCE, TWICE, THREE TIMES AND SO ON.   LAST DATA CARD HAS 99.
C         BEGINNING FROM COLUMN 1.
C
          DIMENSION FREQ(20)
          GAM(Z) = EXP((Z-.5)*ALOG(Z)-Z+.918+(1./(12.*Z))-(1./(360.*Z**3)))
     1    READ (5, 2) FREQ
     2    FORMAT (20F4.0)
          IF (FREQ(1) .EQ. 99.) GO TO 15
C******CALCULATE NEGATIVE BINOMIAL ESTIMATES
          SF = 0.
          SFI = 0.
          SFI2 = 0.
          DO 3 I = 1, 20
          SF = SF + FREQ(I)
          SFI = SFI + FREQ(I) * I
     3    SFI2 = SFI2 + FREQ(I) * I ** 2
          AV1 = SFI/SF
          VAR = (SFI2 - (SFI**2)/SF)/(SF - 1.)
          AV2 = AV1 - ((VAR*FREQ(1))/(AV1*(SF - FREQ(1))))
          TOT = SFI/AV2
          W = (AV1/VAR)*(1.- (FREQ(1)/SF))
          R = ((W * AV1)-(FREQ(1)/SF))/(1.- W)
          WRITE (6, 4) TOT, AV2, W, R
     4    FORMAT  (1H1, 27X, 27HNEGATIVE BINOMIAL ESTIMATES/30H0ESTIMATED
     1POPULATION SIZE = , F9.4/ 33H0MEAN OF COMPLETE DISTRIBUTION = ,
     2 F7.4/ 5H0W = , F9.6/  5H0K = ,F11.6///)
          WRITE (6, 9)
          IF (R .LT. 0.) GO TO 7
          F = 1.
          DO 5 I = 1, 20
          F = F * I
          DIST = SF * W**R*GAM(R+I)*(1.-W)**I / ((1.-W)*GAM(R)*F)
     5    WRITE (6, 6) I, FREQ(I), DIST
     6    FORMAT (1H0, I5I, F7.0, F12.3)
C******CALCULATE GEOMETRIC ESTIMATES
     7    TOT = SF * (SFI - 1.)/(SFI - SF)
          Q  = (SFI - SF)/(SFI - 1.)
          WRITE (6, 8) TOT, Q
     8    FORMAT  (1H1, 27X, 32HGEOMETRIC DISTRIBUTION ESTIMATES/28H0ESTIMAT
     1ED POPULATION SIZE = , F9.4/ 4H0Q = , F6.4)
          WRITE (6, 9)
     9    FORMAT (1H0, 50X, 19HI     DATA     FITTED///)
          DO 10 I = 1, 20
          DIST = TOT * (1. - Q) * Q**I
    10    WRITE (6, 6) I, FREQ(I), DIST
C******CALCULATE POISSON ESTIMATES
          TMEAN = AV1

    11    TMEAN = TMEAN - 0.001
          EST  = TMEAN / (1. - EXP(-TMEAN))
          IF (AV1 - EST) 11, 12, 12
    12    TOT = SFI / TMEAN
          WRITE (6, 13) TOT, TMEAN
    13    FORMAT (1H1, 27X, 17HPOISSON ESTIMATES/29H0ESTIMATED POPULATION
     1SIZE = , F9.4/ 33H0MEAN OF COMPLETE DISTRIBUTION = , F7.4///)
          WRITE (6, 9)
          DIST = TOT * EXP(-TMEAN)
          DO 14 I = 1, 20
          DIST = DIST * TMEAN /. I
    14    WRITE (6, 6) I, FREQ(I), DIST
          GO TO 1
    15    CONTINUE
          STOP
          END
```

Bibliography

Allen, D. L. (1974). Of fire, moose and wolves. *Audubon* **76**: 38–49.

Allen, K. R. (1951). The Horokiwi Stream: a study of a trout population. *New Zeal. Marine Dept. Fish. Bull.*, **No. 10**: 1–231.

Andersen, J. (1953). Analysis of a Danish roe-deer population (*Capreolus capreolus* (L)) based upon the extermination of total stock. *Danish Rev. Game Biol.*, **2**: 127–155.

Andersen, J. (1962). Roe-deer census and population analysis by means of modified marking release technique. pp. 72–80. In Le Cren, E. D., and M. W. Holdgate (eds.) *The Exploitation of Natural Animal Populations*, Blackwell, Oxford.

Andrewartha, H. G., and L. C. Birch (1954). *The Distribution and Abundance of Animals*, University of Chicago Press, Chicago.

Bailey, N. T. J. (1951). On estimating the size of mobile populations from recapture data. *Biometrika*, **38**: 293–306.

Bailey, N. T. J. (1952). Improvements in the interpretation of recapture data. *J. Anim. Ecol.*, **21**: 120–127.

Baker, H. G., and G. L. Stebbins (1965). *The Genetics of Colonizing Species*, Academic Press, New York.

Balham, R. W., and K. H. Miers (1959). Mortality and survival of grey and mallard ducks banded in New Zealand. *New Zeal. Dept. Int. Affairs Wildl. Pubs.*, **No. 5**: 1–56.

Bamford, J. (1970). Evaluating opossum poisoning operations by interference with non-toxic baits. *Proc. New Zeal. Ecol. Soc.*, **17**: 118–125.

Banfield, A. W. F., D. R. Flook, J. P. Kelsall, and A. G. Loughrey (1955). An aerial survey technique for northern big game. *North Amer. Wildl. Conf.*, **20**: 519–532.

Bannikov, A. G. (1950). [Age distributions of a population and its dynamics in *Bombina bombina* L.] *Tr. Akad. Nauk. SSSR*, **70**: 101–103.

Baranov, F. I. (1926). [On the question of the dynamics of the fishing industry.] *Biull. Rybnovo Khoziaistva*, for 1925, **8**: 7–11.

Batcheler, C. L., and D. J. Bell (1970). Experiments in estimating density from joint point—and nearest-neighbour distance samples. *Proc. New Zeal. Ecol. Soc.*, **17**: 111–117.

Batcheler, C. L., J. H. Darwin, and L. T. Pracy (1967). Estimation of opossum (*Trichosurus vulpecula*) populations and results of poison trials from trapping data. *New Zeal. J. Sci.*, **10**: 97–114.

Batcheler, C. L., and P. C. Logan (1963). Assessment of an animal-control campaign in the Harper-Avoca Catchment. *New Zeal. For. Res. Notes*, **27**: 1–27.

Bellrose, F. C. (1955). A comparison of recoveries from reward and standard bands. *J. Wildl. Mgmt.*, **19**: 71–75.

Bellrose, F. C., and E. B. Chase (1950). Population losses in the mallard, black duck and blue-winged teal. *Ill. Nat. Hist. Survey Biol. Notes*, **No. 22**: 1–27.

Bergerud, A. T. (1963). Aerial winter census of caribou. *J. Wildl. Mgmt.*, **27**: 438–449.

Beukema, J. J. (1970). Angling experiments with carp (*Cyprinus carpio* L.) II. Decreasing catchability through one trial learning. *Netherlands J. Zool.*, **20**: 81–92.

Beverton, R. J. H., and S. J. Holt (1957). On the dynamics of exploited fish populations. *Min. Agr. Fish and Food, Fisheries Investigations*, Ser. 2, **19**: 1–533.

Birch, L. C. (1948). The intrinsic rate of natural increase of an insect population. *J. Anim. Ecol.*, **17**: 15–26.

Birch, L. C. (1960). The genetical factor in population ecology. *Amer. Nat.*, **94**: 5–24.

Birch, L. C., Th. Dobzhansky, P. O., Elliott, and R. C. Lewontin (1963). Relative fitness of geographic races of *Drosophila serrata*. *Evolution*, **17**: 72–83.

Bird, J. B. (1970). Paleo-Indian discoidal stones from southern South America. *Amer. Antiquity*, **35**: 205–209.

Blake, C. H. (1951). Wear of towhee bands. *Bird-banding*, **22**: 179–180.

Bodenheimer, F. S. (1938). *Problems of Animal Ecology*, Oxford University Press, London.

Bodenheimer, F. S. (1958). Animal Ecology Today. *Monogr. Biol.*, **6**.

Brand, D. J. (1963). Records of mammals bred in the National Zoological Gardens of South Africa during the period 1908 to 1960. *Proc. Zool. Soc. Lond.*, **140**: 617–659.

Brass, W. (1958). Simplified methods of fitting the truncated negative binomial distribution. *Biometrika*, **45**: 59–68.

Buechner, H. K., I. O. Buss, and H. F. Bryan (1951). Censusing elk by airplane in the Blue Mountains of Washington. *J. Wildl. Mgmt.*, **15**: 81–87.

Buckner, C. H. (1966). The role of vertebrate predators in the biological control of forest insects. *Ann. Rev. Ent.*, **11**: 449–470.

Burnet, F. M. (1962). Evolution made visible: current changes in the pattern of virus disease. pp. 23–32. In Leeper, G. W. (ed.) *The Evolution of Living Organisms*. Melbourne University Press, Melbourne.

Bustard, H. R. (1969). The population ecology of the gekkonid lizard (*Gehyra variegata* (Duméril & Bibron)) in exploited forests in northern New South Wales. *J. Anim. Ecol.*, **38**: 35–51.

Bustard, H. R. (1970). A population study of the scincid lizard *Egernia striolata* in northern New South Wales. *Proc. Koninkl. Nederl. Akad. Wetenschappen*, **C73**: 186–213.

Cassie, R. M. (1954). Some uses of probability paper in the analysis of size frequency distributions. *Austral. J. Mar. Freshw. Res.*, **5**: 513–522.

Cassie, R. M. (1962). Frequency distribution models in the ecology of plankton and other organisms. *J. Anim. Eco!.*, **31**: 65–92.

Caughley, G. (1960). Dead seals inland. *Antarctica*, **2**: 270–271.

Caughley, G. (1963). Dispersal rates of several ungulates introduced into New Zealand. *Nature*, **200**: 280–281.

Caughley, G. (1964). Social organization and daily activity of red kangaroos and grey kangaroos. *J. Mammal.*, **45**: 429–436.

Caughley, G. (1965). Horn rings and tooth eruption as criteria of age in the Himalayan thar, *Hemitragus jemlahicus*. *New Zeal. J. Sci.*, **8**: 333–351.

Caughley, G. (1966). Mortality patterns in mammals. *Ecology*, **47**: 906–918.

Caughely, G. (1967a). Parameters for seasonally breeding populations. *Ecology*, **48**: 834–839.

Caughley, G. (1967b). Calculation of population mortality rate and life expectancy for thar and kangaroos from the ratio of juveniles to adults. *New Zeal. J. Sci.*, **10**: 578–584.

Caughley, G. (1970a). Eruption of ungulate populations, with emphasis on Himalayan thar in New Zealand. *Ecology*, **51**: 54–72.

Caughley, G. (1970b). Liberation, dispersal and distribution of Himalayan thar (*Hemitragus jemlahicus*) in New Zealand. *New Zeal. J. Sci.*, **13**: 220–239.

Caughley, G. (1970c). Population statistics of chamois. *Mammalia*, **34**: 194–199.

Caughley, G. (1971a). Season of births for northern-hemisphere ungulates in New Zealand. *Mammalia*, **35**: 204–219.

Caughley, G. (1971b). Demography, fat reserves and body size of a population of red deer *Cervus elaphus* in New Zealand. *Mammalia*, **35**: 369–383.

Caughley, G. (1974). Bias in aerial survey. *J. Wildl. Mgmt.*, **38**: 921–933.

Caughley, G. (1976). Wildlife management and the dynamics of ungulate populations. *Advances Appl. Biol.*, **1** (in press)

Caughley, G., and L. C. Birch (1971). Rate of increase. *J. Wildl. Magmt.*, **35**: 658–663.

Caughley, G. and J. Caughley (1974). Estimating median date of birth. *J. Wildl. Mgmt.*, **38**: 552–556.

Caughley, G., and J. Goddard (1972). Improving the estimates from inaccurate censuses. *J. Wildl. Mgmt.*, **36**: 135–140.

Caughley, G., R. Sinclair and D. Scott-Kemmis (1976). Experiments in aerial survey. *J. Wildl. Mgmt.*, **40**: 290–300.

Caughley, G., and R. I. Kean (1964). Sex ratios in marsupial pouch young. *Nature*, **204**: 491.

Chapman, D. G. (1951). Some properties of the hypergeometric distribution with applications to zoological sample censuses. *Univ. Calif. Pubs. Stat.*, **1**: 131–160.

Chapman, D. G. (1954). The estimation of biological populations. *Ann. Math. Stats.*, **25**: 1–15.

Chapman, D. G. (1955). Population estimation based on change of composition caused by a selective removal. *Biometrika*, **42**: 279–290.

Chapman, D. G., and G. I. Murphy (1965). Estimates of mortality and population from survey-removal records. *Biometrics*, **21**: 921–935.

Chapman, D. G., and D. S. Robson (1960). The analysis of a catch curve. *Biometrics*, **16**: 354–368.

Child, G., and J. D. Le Riche (1969). Recent springbok treks (mass movements) in south-western Botswana. *Mammalia*, **33**: 499–504.

Child, G., M. B. E. Smith, and W. von Richter (1970). Tsetse control hunting as a measure of large mammal population trends in the Okavango Delta, Botswana. *Mammalia*, **34**: 34–75.

Chittleborough, R. G. (1965). Dynamics of two populations of the humpback whale, *Megaptera novaeangliae* (Borowski). *Austral. J. Mar. Freshw. Res.*, **16**: 33–128.

Christie, A. H. C. (1967). The sensitivity of chamois and red deer to temperature fluctuations. *Proc. New Zeal. Ecol. Soc.*, **14**: 34–39.

Christie, A. H. C., and J. R. H. Andrews (1964). Introduced ungulates in New Zealand. (a) Himalayan thar. *Tuatara*, **12**: 69–77.

Christie, A. H. C., and J. R. H. Andrews (1965). Introduced ungulates in New Zealand. (c) Chamois. *Tuatara*, **13**: 105–111.

Cochran, W. G. (1963). *Sampling Techniques*, John Wiley and Sons, New York. 2nd edition.

Cockrum, E. L. (1948). The distribution of the cotton rat in Kansas. *Trans. Kansas Acad. Sci.*, **51**: 306–312.

Cole, L. C. (1954). The population consequences of life history phenomena. *Quart. Rev. Biol.*, **29**: 103–137.

Cole, L. C. (1957). Sketches of general and comparative demography. *Cold Springs Harb. Symp.*, *Quant. Biol.*, **22**: 1–15.

Cole, L. C. (1960). A note on population parameters in cases of complex reproduction. *Ecology*, **41**: 372–375.

Cormack, R. M. (1966). A test for equal catchability. *Biometrics*, **22**: 330–342.

Cormack, R. M. (1968). The statistics of capture–recapture methods. *Oceanogr. Mar. Biol. Ann. Rev.*, **6**: 455–506.

Coulson, J. C. (1962). The biology of *Tipula subnodicornis* Zetterstedt, with comparative observations on *Tipula paludosa* Meigen. *J. Anim. Ecol.*, **31**: 1–21.

Craig, C. C. (1953). On the utilization of marked specimens in estimating populations of flying insects. *Biometrika*, **40**: 170–176.

220

Cramér, H. (1945). *Mathematical Methods of Statistics*, Princeton University Press, Princeton.

Crissey, W. F. (1967). Aims and methods of waterfowl research in North America. *Trans. 8th Int. Congress Game Biologists:* 37–46.

Dahl, K. (1919). Studies of trout and trout-waters in Norway. *Salmon and Trout Mag.*, **18**: 16–33.

Darroch, J. N. (1958). The multiple recapture census. I. Estimation of a closed population. *Biometrika*, **45**: 343–359.

Darwin, C. (1845). *The Voyage of the Beagle*. Everyman's Library Edition, 1906. Dent and Sons, London

Dasmann, W. P., and R. F. Dasmann (1963). Abundance and scarcity of California deer. *Calif. Fish. and Game*, **49**: 4–15.

Davis, D. E. (1955). Breeding biology of birds. pp. 264–308. In Wolfson, A. (ed.), *Recent Studies in Avian Biology*, University Illinois Press, Urbana.

Davis, D. E. (1956). *Manual for Analysis of Rodent Populations*, Edwards Bros., Ann Arbor.

Davis, D. E., and F. B. Golley (1963). *Principles in Mammalogy*, Reinhold Publishing Corp., New York.

Deevey, E. S. Jr. (1947). Life tables for natural populations of animals. *Quart. Rev. Biol.*, **22**: 283–314.

DeLury, D. B. (1947). On the estimation of biological populations. *Biometrics*, **3**: 145–167.

DeLury, D. B. (1958). The estimation of population size by a marking and recapture procedure. *J. Fish. Res. Bd Canada*, **15**: 19–25.

Douglas, M. H. (1967). Control of thar (*Hemitragus jemlahicus*): evaluation of a poisoning technique. *New Zeal. J. Sci.*, **10**: 511–526.

Durand, D., and J. A. Greenwood (1958). Modifications of the Rayleigh test for uniformity in analysis of two-dimensional orientation data. *J. Geol.*, **66**: 229–238.

Eberhardt, L. L. (1968). A preliminary appraisal of line transects. *J. Wildl. Mgmt.*, **32**: 82–88.

Eberhardt, L. L. (1969a). Population analysis. pp. 457–495. In Giles, R. H. Jr. (ed.). *Wildlife Management Techniques*, 3rd edition revised. Wildlife Society, Washington, D.C.

Eberhardt, L. L. (1969b). Population estimates from recapture frequencies. *J. Wildl. Mgmt.*, **33**: 28–39.

Edwards, R. Y. (1954). Comparison of aerial and ground census of moose. *J. Wildl. Mgmt.*, **18**: 403–404.

Edwards, W. R., and L. L. Eberhardt (1967). Estimating cottontail abundance from live-trapping data. *J. Wildl. Mgmt.*, **31**: 87–96.

Erickson, A. W., and D. B. Siniff (1963). A statistical evaluation of factors influencing aerial survey results. *North Amer. Wildl. Conf.*, **28**: 391–408.

Fabens, A. J. (1965). Properties and fitting of the von Bertalanffy growth curve. *Growth*, **29**: 265–289.

Farner, D. S. (1949). Age groups and longevity in the American robin: comments, further discussion, and certain revisions. *Wilson Bull.*, **61**: 68–81.

Farner, D. S. (1955). Birdbanding in the study of population dynamics. p. 397–449. In Wolfson, A. (ed.), *Recent Studies in Avian Biology*, University Illinois Press, Urbana.

Fenner, F., and F. N. Ratcliffe (1965). *Myxomatosis*, Cambridge University Press, Cambridge.

Filmer, J. F. (1953). Disappointing tests of myxomatosis as rabbit control. *New Zeal. J. Agric.*, **87**: 402–407.

Finney, D. J. (1947). *Probit Analysis: a Statistical Treatment of the Sigmoidal Response Curve*, Cambridge University Press, Cambridge.

Fisher, R. A. (1930). *The Genetical Theory of Natural Selection*, Clarendon Press, Oxford.

Fisher, R. A. (1937). The wave of advance of advantageous genes. *Ann. Eugen.*, **7**: 355–369.

Fisher, R. A. (1941). Note on the efficient fitting of the negative binomial. *Biometrics*, **9**: 197.

Fisher, R. A., and F. Yates (1948). *Statistical Tables for Biological and Medical Research*, 3rd Edition. Oliver and Boyd, Edinburgh.

Ford, E. B. (1964). *Ecological Genetics*. Methuen, London.

Fordham, R. A. (1967). Durability of bands on dominican gulls. *Notornis*, **14**: 28–30.

Frank, P. W. (1960). Prediction of population growth form in *Daphnia pulex* cultures. *Amer. Nat.*, **94**: 357–372.

Frank, P. W. (1968). Life histories and community stability. *Ecology*, **49**: 355–357.

Frith, H. J. (1959). The ecology of wild ducks in inland New South Wales. IV. Breeding. *CSIRO Wildl. Res.*, **4**: 156–181.

Frith, H. J. (1963). Movements and mortality rates of black duck and grey teal in south-eastern Australia. *CSIRO Wildl. Res.*, **8**: 119–131.

Frith, H. J. (1964). Mobility of the red kangaroo *Megaleia rufa*. *CSIRO Wildl. Res.*, **9**: 1–19.

Frith, H. J., and G. B. Sharman (1964). Breeding in wild populations of the red kangaroo, *Megaleia rufa*. *CSIRO Wildl. Res.*, **9**: 86–114.

Fuller, W. A. (1953). Aerial surveys for beaver in the Mackenzie District, Northwest Territories. *North Amer. Wildl. Conf.*, **18**: 329–336.

Geis, A. D., and E. L. Atwood (1961). Proportion of recovered waterfowl bands reported. *J. Wildl. Mgmt.*, **25**: 154–159.

Gentry, J. B. (1968). Dynamics of an enclosed population of pine mice, *Microtus pine-torum*. *Res. Popul. Ecol.*, **10**: 21–30.

Gerking, S. D. (1967). *The Biological Basis of Freshwater Fish Production*, Blackwell, London.

Gerrard, D. J., and H. C. Chiang (1970). Density estimation of corn rootworm egg populations based upon frequency of occurrence. *Ecology*, **51**: 237–245.

Getz, L. L. (1961). Response of small mammals to live-traps and weather conditions. *Amer. Midl. Nat.*, **66**: 160–170.

Gilbert, P. F., and J. R. Grieb (1957). Comparison of air and ground deer counts in Colorado. *J. Wildl. Mgmt.*, **21**: 33–37.

Giles, R. H. Jr. (1969). *Wildlife Management Techniques*, 3rd edition revised. The Wildlife Society, Washington, D.C.

Goddard, J. (1967). The validity of censusing black rhinoceros populations from the air. *East African Wildl. J.*, **5**: 18–23.

Goin, C. J., and O. B. Goin (1962). *Introduction to Herpetology*, Freeman, San Francisco.

Goodman, L. A. (1967). On the reconciliation of mathematical theories of population growth. *J. Roy. Stat. Soc.*, **A130**: 541–553.

Goodman, L. A. (1968). An elementary approach to the population projection-matrix and to the mathematical theory of population growth. *Demography*, **5**: 382–409.

Graham, M. (1935). Modern theory of exploiting a fishery, and application to North Sea trawling. *J. Conseil Expl. Mer.*, **10**: 264–274.

Green, N. B. (1957). A study of the life history of *Pseudacris brachyphona* (Cope) in West Virginia with special reference to behaviour and growth of marked individuals. *Dissertation Abstr.*, **17**: 23692.

Gulland, J. A. (1955). Estimation of growth and mortality in commercial fish populations. *U.K. Min. Agric. Fish.; Fish Invest. Series 2*, **8**: 1–46.

Gulland, J. A. (1962). The application of mathematical models to fish populations. pp. 204–217. In E. D. Le Cren and M. W. Holdgate (eds.) *The Exploitation of Natural Animal Populations*, Blackwell, Oxford.

Gulland, J. A. (1968). Manual of methods for fish assessment. Pt. I. Fish population analysis. *FAO Fish. Tech. Paper*, **No. 40,** Revision 2.

Hairston, N. G., D. W. Tinkle, and H. M. Wilbur (1970). Natural selection and the parameters of population growth. *J. Wildl. Mgmt.*, **34**: 681–690.

Hanson, W. R. (1963). Calculation of productivity, survival and abundance of selected vertebrates from sex and age ratios. *Wildl. Monog.*, No. 9.

Hanson, W. R. (1967). Estimating the density of an animal population. *J. Res. Lepidoptera*, **6**: 203–247.

Harte, J. L. (1932). Statistics of the whitefish (*Coregonus clupeaformis*) population of Shakespeare Island Lake, Ontario. *Univ. Toronto Studies, Biol. Ser.*, **No. 36**: 1–28.

Hickey, F. (1960). Death and reproductive rate in relation to flock culling and selection. *New Zealand J. Agric. Res.*, **3**: 332–344.

Hickey, F. (1963). Sheep mortality in New Zealand. *New Zealand Agriculturalist*, **15**: 1–3.

Hickey, J. J. (1952). Survival studies of banded birds. *U.S. Fish. Wildl. Serv., Spec. Sci. Rept., Wildl.*, **15**: 1–177.

Hjort, J., G. Jahn, and P. Ottestad (1933). The optimum catch. *Hvalradets Skrifter*, **7**: 92–127.

Höglund, N., G. Nilsson, and F. Stålfelt (1967). Analysis of a technique for estimating willow grouse (*Lagopus lagopus*) density. *Trans. 8th Int. Cong. Game Biologists:* 156–159.

Holgate, P. (1966). Contributions to the mathematics of animal trapping. *Biometrics*, **22**: 925–936.

Holgate, P. (1967). Population survival and life history phenomena. *J. Theoret. Biol.*, **14**: 1–10.

Howard, W. E. (1959). The rabbit problem in New Zealand. *New Zeal. Dept. Sci. Indust. Res., Info. Ser.*, **No. 16**: 1–47.

Howard, W. E. (1960). Innate and environmental dispersal of individual vertebrates. *Amer. Midland Nat.*, **63**: 152–161.

Howard, W. E., and H. E. Childs, Jr. (1959). Ecology of pocket gophers with emphasis on *Thomomys bottae mewa*. *Hilgardia*, **29**: 277–358.

Humphrey, S. R. (1974). Zoogeography of the nine-banded armadillo (*Dasypus novemcinctus*) in the United States. *Bioscience*, **24**: 457–462.

Irwin, J. O. (1959). On the estimation of the mean of a Poisson distribution from a sample with the zero class missing. *Biometrics*, **15**: 324–326.

Jenkins, D., and A. Watson. (1962). Fluctuations in a red grouse (*Lagopus scoticus* (Latham) population, 1956–9. pp. 96–130. In E. D. Le Cren and M. W. Holdgate (eds) *The Exploitation of Natural Animal Populations*. Blackwell, Oxford.

Johnson, D. H. (1974). Estimating survival rates from banding of adult and juvenile birds. *J. Wildl. Mgmt.*, **38**: 290–297.

Jolly, G. M. (1965). Explicit estimates from capture–recapture data with both death and immigration—stochastic model. *Biometrika*, **52**: 225–247.

Jolly, G. M. (1969). Sampling methods for aerial censuses of wildlife populations. *East African Agr. For. J.*, **34**: 46–49.

Kalela, O. (1949). Changes in geographic ranges in the avifauna of northern and central Europe in relation to recent changes in climate. *Bird-banding*, **20**: 77–103.

Kelker, G. H. (1940). Estimating deer populations by a differential hunting loss in the sexes. *Proc. Utah Acad. Sci. Arts and Letters*, **17**: 65–69.

Kelker, G. H. (1944). Sex ratio equations and formulas for determining wildlife populations. *Proc. Utah Acad. Sci. Arts and Letters*, **20**: 189–198.

Kelker, G. H. (1947). Computing the rate of increase for deer. *J. Wildl. Mgmt.*, **11**: 177–183.

Kelker, G. H. (1949–50). Sex and age class ratios among vertebrate populations. *Proc. Utah Acad. Sci. Arts and Letters*, **27**: 12–21.

Kelker, G. H. (1952). Yield tables for big game herds. *J. Forestry*, **50**: 206–207.

Kelker, G. H., and W. R. Hanson. (1964). Simplifying the calculation of differential survival of age-classes. *J. Wildl. Mgmt.*, **28**: 411.

Kendall, D. G. (1948). A form of wave propagation associated with the equation of heat conduction. *Proc. Camb. Phil. Soc.*, **44**: 591–594.

223

Kennedy, W. A. (1951). The relationship of fishing effort by gill nets to interval between lifts. *J. Fish. Res. Bd. Canada*, **8**: 264–274.

Kennedy, W. A. (1953). Growth, maturity, fecundity and mortality in relatively unexploited whitefish, *Coregonus clupeaformis*, of Great Slave Lake. *J. Fish. Res. Bd. Canada*, **10**: 413–441.

Kennedy, W. A. (1954). Growth, maturity and mortality in the relatively unexploited lake trout, *Cristivomer namaycush*, of Great Slave Lake. *J. Fish. Res. Bd. Canada*, **11**: 827–852.

Kikkawa, J. (1964). Movement, activity and distribution of the small rodents *Clethrionomys glareolus* and *Apodemus sylvaticus* in woodland. *J. Anim. Ecol.*, **33**: 259–299.

Kirkpatrick, T. H. (1965). Studies of the Macropodidae in Queensland. 3. Reproduction in the grey kangaroo. *Qd. J. Agric. Anim. Sci.*, **22**: 319–328.

Klomp, H., M. A. J. van Montfort, and P. M. L. Tammes (1964). Sexual reproduction and underpopulation. *Arch. Néerlandaises Zool.*, **16**: 105–110.

Klott, E. (1965). *Factors Affecting Estimates of Meadow Mouse Populations*, Ph.D. Thesis, University of Toronto.

Kraak, W. G., G. L. Rinkel, and E. Hoogerheide (1940). Oecologische bewerking van de Europese ringgegevens van de kievit (*Vanellus vanellus* (L.)). *Ardea*, **29**: 151–175.

Lack, D. (1943a). *The Life of the Robin*, H. F. and G. Witherby, London.

Lack, D. (1943b). The age of the blackbird. *Brit. Birds*, **36**: 166–175.

Lack, D. (1943c). The age of some more British birds, *Brit. Birds:* 193–197, 214–221.

Lack, D. (1966). *Population Studies of Birds*. Clarendon Press, Oxford.

Lamprey, H. F. (1964). Estimation of large mammal densities, biomass and energy exchange in the Tarangire Game Reserve and the Masai Steppe in Tanganyika. *East African Wildl. J.*, **2**: 1–46.

Lander, R. H. (1962). A method of estimating mortality rates from a change of composition. *J. Fish. Res. Bd. Canada*, **19**: 159–168.

Laughlin, R. (1965). Capacity for increase: a useful population statistic. *J. Anim. Ecol.*, **34**: 77–91.

Lefkovitch, L. P. (1966). A population growth model incorporating delayed responses. *Bull. Math. Biophys.*, **28**: 219–233.

Lefkovitch, L. P. (1967). A theoretical evaluation of population growth after removing individuals from some age classes. *Bull. Ent. Res.*, **57**: 437–445.

LeResche, R. E., and R. A. Rausch (1974). Accuracy and precision of aerial moose counting. *J. Wildl. Mgmt.*, **38**: 175–182.

Leslie, P. H. (1948). Some further notes on the use of matrices in population analysis. *Biometrika*, **35:** 213–245.

Leslie, P. H. (1952). The estimation of population parameters from data obtained by means of the capture-recapture method. II. The estimation of total numbers. *Biometrika*, **39**: 363–388.

Leslie, P. H. (1959). The properties of a certain lag type of population growth and the influence of an external lag factor on a number of such populations. *Physiol. Zoöl.* **37**: 151–159.

Leslie, P. H. (1966). The intrinsic rate of increase and overlap of successive generations in a population of guillemots (*Uria aalge* Pont.). *J. Anim. Ecol.*, **35**: 291–301.

Leslie, P. H., and D. Chitty (1951). The estimation of population parameters from data obtained by means of the capture–recapture method. I. The maximum likelihood equations for estimating the death-rate. *Biometrika*, **38**: 269–292.

Leslie, P. H., D. Chitty, and H. Chitty (1953). The estimation of population parameters from data obtained by means of the capture-recapture method. III. An example of the practical applications of the method. *Biometrika*, **40**: 137–169.

Leslie, P. H., and D. H. S. Davis (1939). An attempt to determine the absolute number of rats on a given area. *J. Anim. Ecol.*, **8**: 94–113.

Leslie, P. H., and R. M. Ranson (1940). The mortality, fertility and rate of natural increase of the vole (*Microtus agrestis*) as observed in the laboratory. *J. Anim. Ecol.*, **9**: 27–52.

Leslie, P. H., T. S. Tener, M. Vizoso, and H. Chitty (1955). The longevity and fertility of the Orkney vole, *Microtus orcadensis*, as observed in the laboratory. *Proc. Zool. Soc. Lond.*, **125**: 115–125.

Leslie, P. H., U. M. Venables, and L. S. V. Venables (1952). The fertility and population structure of the brown rat (*Rattus norvegicus*) in corn-ricks and some other habitats. *Proc. Zool. Soc. Lond.*, **122**: 187–238.

Lewis, E. G. (1942). On the generation and growth of a population. *Sankhya*, **6**: 93–96.

Lewontin, R. C. (1965). Selection for colonizing ability. pp. 77–91. In Baker, H. G., and G. L. Stebbins (eds). *The Genetics of Colonizing Species*. Academic Press, New York.

Lincoln, F. C. (1930). Calculating waterfowl abundance on the basis of banding returns. *Cir. U.S. Dept. Agric.*, No. 118.

Lister, R. R. (1969). *The Secret History of Genghis Khan*, Peter Davies, London.

Lotka, A. J. (1907a). Relationship between birth rates and death rates. *Science*, **26**: 21–22.

Lotka, A. J. (1907b). Studies on the mode of growth of material aggregates. *Amer. J. Sci.*, 4th Series, **24**: 199–216.

Lovell, H. B. (1948). The removal of bands by cardinals. *Bird-banding* **19**: 70–71.

Low, W. A., and I. McT. Cowan (1963). Age determination of deer by annular structure of dental cementum. *J. Wildl. Mgmt.*, **27**: 466–471.

Lowe, V. P. W. (1969). Population dynamics of the red deer (*Cervus elaphus* L.) on Rhum. *J. Anim. Ecol.*, **38**: 425–457.

MacArthur, R. (1968). Selection for life tables in periodic environments. *Amer. Nat.*, **102**: 381–383.

MacLeod, J. (1958). The estimation of numbers of mobile insects from low-incidence recapture data. *Trans. Roy. Ent. Soc. Lond.*, **110**: 363–392.

Marten, G. G. (1970). A regression method for mark-recapture estimation of population size with unequal catchability. *Ecology*, **51**: 291–295.

Martin, P. S. (1973). The discovery of America. *Science*, **179**: 969–974.

May, R. M. (1973). *Stability and Complexity in Model Ecosystems*, Princeton University Press, Princeton.

Meats, A. (1971). The relative importance to population increase of fluctuations in mortality, fecundity and the time variables of the reproductive schedule. *Oecologia*, **6**: 223–237.

Mech, L. D. (1966). The wolves of Isle Royale. *Fauna Nat. Parks U.S.A. Fauna Series*, **7**: 1–201.

Miller, A. H. (1962). Bimodal occurrence of breeding in an equatorial sparrow. *Proc. Nat. Acad. Sci. U.S.*, **48**: 396–400.

Moran, P. A. P. (1951). A mathematical theory of animal trapping. *Biometrika*, **38**: 307–311.

Morisita, M. (1965). The fitting of the logistic equation to the rate of increase of population density. *Res. Popul. Ecol.*, **7**: 52–55.

Morris, R. F. (1957). The interpretation of mortality data in studies on population dynamics. *Canad. Ent.*, **89**: 49–69.

Morris, R. F. (1963). The dynamics of epidemic spruce budworm populations. *Mem. Ent. Soc. Canada*, **31**: 1–332.

Murdoch, W. W. (1966). Population stability and life history phenomena. *Amer. Nat.*, **100**: 5–11.

Murie, A. (1944). The wolves of Mount McKinley. *Fauna Nat. Parks U.S.*, *Fauna Series*, **5**.

Murphy, D. A. (1963). A captive elk herd in Missouri, *J. Wildl. Mgmt.*, **27**: 411–414.

Murphy, G. I. (1966). Population dynamics and population estimation. In A. Calhoun (ed.) *Inland Fisheries Management*, Calif. Dept. of Fish and Game, San Francisco.

Murphy, G. I. (1967). Vital statistics of the Pacific sardine (*Sardinops caerulea*) and the population consequences. *Ecology*, **48**: 731–736.

Murphy, G. I. (1968). Pattern in life history and the environment. *Amer. Nat.*, **102**: 391–403.

Myers, J. H., and C. J. Krebs (1971). Genetic, behavioral and reproductive attributes of dispersing field voles *Microtus pennsylvanicus* and *Microtus ochrogaster*. *Ecol. Monog.*, **41**: 53–78.

Myers, K. (1970). The rabbit in Australia. pp.478–506. In P. J. den Boer and G. R. Gradwell, eds: *Dynamics of Populations*. Adv. Study Inst. Dynamics Numbers Popul. (Oosterbeck).

Nagel, W. P., and D. Pimentel (1964). The intrinsic rate of natural increase of the ptero-malid parasite *Nasonia vitripennis* (Walk.) on its muscoid host *Musca domestica* L. *Ecology*, **45**: 658–660.

Neyman, J. (1934). On the two different aspects of the representative method: the method of statified sampling and the method of purposive selection. *J. Roy. Stat. Soc.*, **97**: 558–606.

Nice, M. M. (1937). Studies on the life history of the song sparrow. I. A population study of the song sparrow. *Trans. Linn. Soc. N.Y.*, **4**: 1–247.

Niven, B. S. (1970). Mathematics of populations of the quokka, *Setonix brachyurus* (Marsupialia). I. A simple deterministic model for quokka populations. *Austral. J. Zool.*, **18**: 209–214.

Organ, J. A. (1961). Studies of the local distribution, life history, and population dynamics of the salamander genus *Desmognathus* in Virginia. *Ecol. Monog.*, **31**: 189–220.

Orians, G. H., and P. H. Leslie (1958). A capture–recapture analysis of a shearwater population. *J. Anim. Ecol.*, **27**: 71–86.

Parr, M. J., T. J. Gaskell, and B. J. George (1968). Capture–recapture methods of esti-mating animal numbers. *J. Biol. Educ.*, **2**: 95–117.

Peabody, F. E. (1961). Annual growth zones in living and fossil vertebrates. *J. Morph.*, **108**: 11–62.

Petersen, C. G. J. (1896). The yearly immigration of young plaice into Limfjord from the German sea, etc. *Rept. Danish Biol. Stn.*, **6**: 1–48.

Petrides, G. A. (1949). Viewpoints on the analysis of open season sex and age ratios. *Trans. N. Am. Wildl. Conf.*, **14**: 391–410.

Petrides, G. A. (1953). Aerial deer counts. *J. Wildl. Mgmt.*, **17**: 97–98.

Petrides, G. A. (1954). Estimating the percentage kill in ring-necked pheasants and other game species. *J. Wildl. Mgmt.*, **18**: 294–297.

Phillips, B. F., and N. A. Campbell (1970). Comparison of methods of estimating popula-tion size using data on the whelk *Dicathais aegrota* (Reeve). *J. Anim. Ecol.*, **39**: 753–759.

Pielou, E. C. (1969). *An Introduction to Mathematical Ecology*, Wiley-Interscience, New York.

Pielou, E. C. (1974). *Population and Community Ecology*. Gordon and Breach, New York.

Poole, W. E. (1973). A study of breeding in the grey kangaroo *Macropus giganteus* Shaw and *M. fuliginosus* (Desmarest), in central New South Wales. *Aust. J. Zool.* **21**: 183–212.

Rasmussen, D. I. (1941). Biotic communities of Kaibab Plateau, Arizona. *Ecol. Monogr*, **3**: 229–275.

Reddingius, J. and P. J. den Boer (1970). Simulation experiments illustrating stabilization of animal numbers by spreading the risk. *Oecologia*, **5**: 240–284.

Richdale, L. E. (1954). *A Population Study of Penguins*, Oxford University Press, London.

Ricker, W. E. (1940). Relation of 'catch per unit effort' to abundance and rate of exploita-tion. *J. Fish. Res. Bd. Canada*, **5**: 43–70.

Ricker, W. E. (1949). Mortality rates in some little-exploited populations of freshwater fish. *Trans. Am. Fish. Soc.*, for 1947, **77**: 114–128.

Ricker, W. E. (1954a). Stock and recruitment. *J. Fish Res. Bd. Canada*, **11**: 559–623.

Ricker, W. E. (1954b). Effects of compensatory mortality upon population abundance. *J. Wildl. Mgmt.*, **18**: 45–51.

Ricker, W. E. (1958). Handbook of computations for biological statistics of fish populations. *Fish. Res. Bd. Canada Bull.*, **No. 119**: 1–300.

Riffenburgh, R. H. (1969). A stochastic model of interpopulation dynamics in marine ecology. *J. Fish Res. Bd. Canada*, **26**: 2843–2880.

Riney, T. (1956a). Differences in proportion of fawns to hinds in red deer (*Cervus elaphus*) from several New Zealand environments. *Nature*, **177**: 488–489.

Riney, T. (1956b). Comparison of occurrence of introduced animals with critical conservation areas to determine priorities for control. *New Zeal. J. Sci. Tech.*, **B38**: 1–18.

Riney, T. (1957). The use of faeces counts in studies of several free-ranging mammals in New Zealand. *New Zeal. J. Sci. Tech.*, **B38**: 507–522.

Robinette, W. L. (1949). Winter mortality among mule deer in Fishlake National Forest, Utah. *U.S. Fish. and Wildl. Serv., Spec. Sci. Report*, **65**: 1–15.

Robinette, W. L., D. A. Jones, G. Rogers, and J. S. Gashwiler (1957). Notes on tooth development and wear for Rocky Mountain mule deer. *J. Wildl. Mgmt.*, **21**: 134–153.

Roff, D. A. (1973). An examination of some statistical tests used in the analysis of mark-recapture data. *Oecologia* **12**: 35–54.

Rogers, A. (1968). *Matrix Analysis of Interregional Population Growth and Distribution*. University of California Press, Berkeley.

Rounsefell, G. A., and W. H. Everhart (1953). *Fisheries Science: Its Methods and Applications*, John Wiley and Sons, New York.

Rowley, I. (1966). Rapid band wear on Australian ravens. *Austral. Bird Bander*, **4**: 47–49.

Rupp, R. S. (1966). Generalized equation for the ratio method of estimating population abundance. *J. Wildl. Mgmt.*, **30**: 523–526.

Ryel, L. A., L. D. Fay, and R. C. van Etten (1961). Validity of age determination in Michigan deer. *Pap. Mich. Acad. Sci.*, **46**: 289–316.

Sampford, M. R. (1955). The truncated negative binomial distribution. *Biometrika*, **42**: 58–69.

Schaefer, M. B. (1957). A study of the dynamics of the fishery for yellowfin tuna in the eastern tropical Pacific Ocean. *Inter-American Trop. Tuna Comm. Bull.*, **2**: 247–267.

Schnabel, Z. E. (1938). The estimation of total fish in a lake. *Am. Math. Mon.*, **45**: 348–352.

Schumacher, F. X., and R. W. Eschmeyer (1943). The estimation of fish populations in lakes and ponds. *J. Tenn. Acad. Sci.*, **18**: 228–249.

Seber, G. A. F. (1965). A note on the multiple-recapture census. *Biometrika*, **52**: 249–259.

Seber, G. A. F. (1970). Estimating time-specific survival and reporting rates for adult birds from band returns. *Biometrika*, **57**: 313–318.

Seber, G. A. F. (1972). Estimating survival rates from bird-band returns. *J. Wildl. Mgmt.* **36**: 405–413.

Seber, G. A. F. (1973). *The Estimation of Animal Abundance and Related Parameters*. Griffin, London.

Seber, G. A. F., and E. D. Le Cren (1967). Estimating population parameters from catches large relative to the population. *J. Anim. Ecol.*, **36**: 631–643.

Selleck, D. M., and C. M. Hart (1957). Calculating the percentage of kill from sex and age ratios. *Calif. Fish and Game*, **43**: 309–316.

Sharpe, F. R., and A. J. Lotka (1911). A problem in age-distribution. *Phil. Mag.*, **21**: 435–438.

Silliman, R. P., and J. S. Gutsell (1958). Experimental exploitation of fish populations. *Fish. Bull., Fish Wildl. Serv.*, **58**: 215–252.

Simpson, G. G., A. Roe, and R. C. Lewontin (1960). *Quantitative Zoology*, Revised edition. Harcourt, Brace and Company, New York.

Siniff, D. B., and R. O. Skoog (1964). Aerial censusing of caribou by stratified random sampling. *J. Wildl. Mgmt.*, **28**: 391–401.

Skellam, J. G. (1951). Random dispersal in theoretical populations. *Biometrika*, **38**: 196–218.

Skellam, J. G. (1955). The mathematical approach to population dynamics. pp. 31–45.

In Cragg, J. B. and N. W. Pirie (eds). *The Numbers of Man and Other Animals*, Oliver and Boyd, Edinburgh.

Snedecor, G. W., and W. G. Cochran (1967). *Statistical Methods*, 6th edition. Iowa State University Press, Ames.

Sonleitner, F. J., and M. A. Bateman (1963). Mark–recapture analysis of a population of Queensland fruit-fly, *Dacus tryoni* (Frogg.) *J. Anim. Ecol.*, **32**: 259–269.

Southern, H. N. (1964). *The Handbook of British Mammals*, Blackwell, Oxford.

Southwood, T. R. E. (1966). *Ecological Methods with Particular Reference to the Study of Insect Populations*, Methuen, London.

Stewart, R. E., A. D. Geis, and C. D. Evans (1958). Distribution of populations and hunting kill of the canvasback. *J. Wildl. Mgmt.*, **22**: 333–370.

Stoddart, D. M. (1950). Individual range, dispersion and dispersal of water voles (*Avicola terrestris* (L.)). *J. Anim. Ecol.*, **39**: 403–425.

Taber, R. D. (1969). Criteria of sex and age. pp. 325–461. In Giles, R. H. (ed.) *Wildlife management Techniques*, The Wildlife Society, Washington, D. C. 3rd edn, revised.

Talbot, L. M., and D. R. M. Stewart (1964). First wildlife census of the entire Serengeti-Mara region, East Africa. *J. Wildl. Mgmt.*, **28**: 815–827.

Tanton, M. T. (1965). Problems of live trapping and population estimation of the wood mouse *Apodemus sylvaticus* (L.) *J. Anim. Ecol.*, **34**: 1–22.

Tanton, M. T. (1969). The estimation and biology of the bank vole (*Clethrionomys glareolus* (Schr.)) and wood mouse (*Apodemus sylvaticus* (L.)). *J. Anim. Ecol.*, **38**: 511–529.

Tast, J. (1966). The root vole *Microtus oeconomus* (Pallas), an inhabitant of seasonally flooded land. *Ann. Zool. Fenn.*, **3**: 127–171.

Taylor, B. J. R. (1965). The analysis of polymodal frequency distributions. *J. Anim. Ecol.*, **34**: 445–452.

Taylor, S. M. (1966). Recent quantitative work on British bird populations. A review. *The Statistician*, **16**: 119–170.

Tinkle, D. W. (1967). The life and demography of the side-blotched lizard, *Uta stansburiana*. *Misc. Publ. Mus. Zool. Univ. Mich.*, **132**: 1–182.

Tinkle, D. W. (1973). A population analysis of the sagebrush lizard, *Sceloporus graciosus*, in Southern Utah. *Copeia*, **1973**: 284–296.

Tomlinson, R. E. (1968). Reward banding to determine reporting rates of recovered mourning dove bands. *J. Wildl. Mgmt.*, **32**: 6–11.

Tschuprow, A. A. (1923). On the mathematical expectation of the moment frequency distributions in the case of correlated observations. *Metron*, **2**: 461–493, 646–683.

Turner, F. B. (1960a). Tests of randomness in recapture of *Rana p. pretiosa. Ecology*, **41**: 237–239.

Turner, F. B. (1960b). Size and dispersion of a Louisiana population of the cricket frog, *Acris gryllus. Ecology*, **41**: 258–268.

Turner, F. B. (1962). The demography of frogs and toads. *Quart. Rev. Biol.*, **37**: 303–314.

Watson, R. M. (1969). Aerial photographic methods in censuses of animals. *East African Agric. For. J.*, **34**: 32–37.

Watson, R. M., G. H. Freeman, and G. M. Jolly (1969). Some indoor experiments to simulate problems in aerial censusing. *East. African Agric. For. J.*, **34**: 56–59.

Watson, R. M., G. M. Jolly, and A. D. Graham (1969). Two experimental censuses. *East African Agric. For. J.*, **34**: 60–62.

Watson, R. M., I. S. C. Parker, and T. Allan (1969). A census of elephants and other large mammals in the Mkomazi region of northern Tanzania and Southern Kenya. *E. Afr. Wildl. J.*, **7**: 11–26.

White, E. G. (1971). A computer program for capture–recapture studies of animal populations: a Fortran listing for the stochastic model of G. M. Jolly. *New Zeal. Tussock Grasslands and Mountain Lands Inst. Spec. Pub.*, **8**: 1–33.

W. H. O. (1972). *World Health Statistics Annual for 1969. I.* W. H. O., Geneva.

Wiener, G., and H. P. Donald (1955). A study of variation in twin cattle. IV. Emergence of permanent incisor teeth. *J. Dairy Res.*, **22**: 127–137.

Wohlschlag, D. E. (1954). Mortality rates of whitefish in an Arctic lake. *Ecology*, **35**: 388–396.

Wolfson, A. (1955). *Recent Studies in Avian Biology*, University of Illinois Press, Urbana.

Zippin, C. (1956). An evaluation of the removal method of estimating animal populations. *Biometrics*, **12**: 163–189.

Author Index

Subject Index